A Complete Guide
to Postpartum Recovery

徹底改善體質，終身受用

坐好月子

目錄

第五章

第六章

很高興與圓方出版社合作製作《坐好月子》，這本新書有別於筆者過往的食療書籍，其特色是除了着重產後補身食療外，還提供了很多有關產時、產後一些在過往不為人所重視的資訊，也解釋了很多關於坐月習俗的迷思，務求消除一般人以為中醫欠缺科學根據的想法，希望讀者可以從這本書的內容，明白到中醫是如何看待坐月的。

　　筆者在行醫及教授陪月課程的時候，時常被病人或學生告知，西醫在產婦出院時會有一堆要忌口的食物及藥材（例如：木耳、紅棗、雲耳、生化湯、酒、薑、醋等），告誡產婦不准食用，但這些食材又是中醫及中國民間在坐月期間最常讓產婦吃的，到底應該聽從哪方面的意見好呢？當碰到這種兩難的局面時，筆者最經常的做法是：先了解西醫到底是基於甚麼道理去禁止產婦食用某些食材，而該產婦的體質又是否有特殊狀況？再以中醫的角度去衡量產婦是否不適合服用這些食材。舉例：一個身體狀況很正常的產婦，其生產過程又沒有異常之處，產後也沒有任何不妥，為甚麼不可適量進食薑、木耳、煮過的雞酒、紅棗、醋呢？請注意：是**適量地**進食，不是盲目地吃，天天吃、餐餐吃。

　　那些被西醫禁止食用的食材，無疑或多或少均具有活血化瘀的作用，但中醫的活血化瘀往往不能簡單的視之為增加出血傾向及令人流血過多，中醫有「活血止血」這一種治療方法，活血有時反而可有效止血或減少流血；化瘀也不一定是要令產婦惡露無限或不正常地增多，化瘀往往是為了幫助子宮收縮而令惡露盡快排淨，減少腹痛。其目的是為了令產婦在產後盡快恢復健康，並維持身體在相對良好的狀況，以利哺乳及照顧嬰兒。西醫時常會說，若吃了那些具有活血化瘀功效的食材，產婦就容易出意外、易大出血及流血不止。筆者也不排除某些人是不宜進食太多活血、補血、化瘀食／藥材的；但絕對不同意但凡是吃了這些食／藥材均會令人出血或失血。

　　回憶起筆者還是助產士及護士時的經歷：有一晚病房收了一個急性產後大出血的病人，該產婦是順產第 5 天後突然在家暈倒，下體流出大量鮮血而被送急診的，要緊急輸血及用止血藥，在病房留院治療 3 天後，下體出血減少至正常惡露量。據了解，產婦的家姑每天均煮大量木耳煮雞酒及在藥材舖自行購買活血、行血化瘀的藥材煮湯給該病人喝，而該病人的凝血（即正常出血後可自行止血）時間相對一般人是較為長的。醫生估計她的出血是因為不當服用中藥引起。現在回想起來，筆者也同意醫生當時的判斷，但不代表木耳煮雞酒必然會造成所有產婦產後大出血。要結合幾方面去考慮，例如：所服用的木耳份量及頻密度？酒的份量及何種酒？煮的時間是否足以把酒精揮發掉？產婦的體質是否適合服用？

至於其他藥材，則更是不能一概而論。沒有經過中醫師診斷及處方而服用中藥材是危險的，這也正如沒經西醫診斷而自行購買西藥服食一樣，風險相等。出了事，當然不能怪是中藥的錯。

筆者只能奉勸讀者，切勿以為中藥材如當歸、川芎、田七等是很普通的藥材，而不需要經中醫師診治便可自行服用，產後最好是向中醫師諮詢，判斷身體是否適合才服用補身藥材。中醫師之所以要取得專業資格才能行醫，就是為了確保經他／她們診治的病者有足夠的保障，中醫師會根據個別產婦的體質來處方適合的藥材給她們補身，即使是傳統的生化湯也並非不按病者體質而千篇一律的去用藥及組方的。所以產婦們可以放心服用經醫師處方的食／藥材。

再跟讀者分享一個病案，這個病人是筆者一位相熟病人的妹妹，她很幸運，因為她就住在某醫院附近，因而救回一命。她的故事是：剖腹生產後第 5 天，突然急性大出血而休克昏迷，被緊急送往家門對面的醫院進行剖腹探查術止血，她更幸運的是不用把子宮切除也能夠止血（因為在緊急情況下為了保住病人的性命，醫生或會把子宮切掉為病人止血救命）。筆者詳細詢問病人到底西醫有沒有找出是甚麼原因導致她突然內出血？病人說西醫也未能確實解釋，只是有詢問她服用了甚麼中藥及食療；由於該病人一向遵從西醫的吩咐，因此，在產後到出事期間，完全沒有進食過任何藥材或西醫禁止的食物，所以西醫說是不明原因的出血。這個病人的經歷雖只是單一的例子，但可反映出有些意外、病況的發生是跟吃了甚麼沒有必然關係的。很奇怪有些人減肥會找營養師諮詢，但產後補身卻不會諮詢中醫師而自信自己可以完全掌握中藥材的功效而自行服用，到出事時即去看西醫，一般西醫並不完全了解中醫藥的理論及概念，難免會歸咎於中藥甚至中醫，「一刀切」地禁止病人服食中藥甚至是某些普通食材。筆者嘗試站在那些西醫的角度看，囑咐患者不進食、不使用中藥材當然是最為安全的做法，既然他／她們不懂中醫，總之不吃就最安全，所以才會出現這些矛盾的現象。

筆者不是想批判甚麼，只是想指出為了盡量降低錯服補身食材藥材而出事的風險，產婦最好在產後先找相熟中醫師診證，之後按醫師指示去進行食療補身才是最佳做法，中醫師所受的教育及訓練就像西醫一樣，是用所學到的專業知識來幫助患者治療疾病及強身健體的，胡亂進補，不單對產婦有危害，對所有人也同樣有風險。希望讀者可以通過書中的分享，學懂、學對一些關於產後安全補健養生的「小貼士」，調養好身體，養育出壯健的小寶寶。

歡迎讀者傳電郵至 info@chimed.com.hk 來分享你們的經歷。

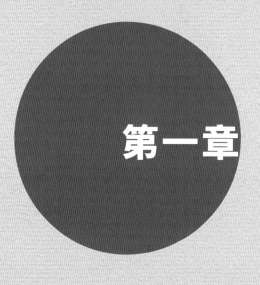

第一章

由於產房與手術室室溫較低，產婦在待產及手術期間容易着涼，如何可以避免？筆者以過往助產士的見聞來介紹產房保暖或穿衣之道，希望各位準媽咪都能健健康康地迎接小生命的來臨。

產房與手術室之保暖攻略

　　筆者在未行醫之前曾是一位助產士及護士，在大型綜合醫院的不同部門工作了十年，以在產房工作的印象最難忘。醫院對大多數人來說，是悲多於喜的地方，唯獨產房的門內門外，你總會看到一張張充滿期待，然後展露喜悅的面孔，因為對現代人來說，醫院產房就等同生命降臨之地，是大部分新生命墜入紅塵的初始之地。

　　20 年前筆者工作的產房，環境是很「冷酷」及「嚴肅」的，不親切也不算「友善」。此話怎講？因為從前仍不太鼓勵丈夫陪產，而環境的設計也令一般人覺得很「嚴肅」，工作人員都很「正經」，因而令人有一種「冷酷」的感覺，令到很多產婦產生一種「驚」的錯覺。現在的產房設計當然是多了很多「親和、溫柔」的元素，亦鼓勵丈夫陪產，使夫婦可以共同面對生產時的痛苦，一同分享迎接新生命的喜悅，減少產婦孤身一人在陌生環境「搏鬥」的恐懼，也舒緩了產婦的緊張不安情緒，令生產過程更順利。

　　不過，根據個人經驗及很多病人的反映，今時今日的產房環境雖然改善了很多，也改良了某些常規程序，讓人感覺親和、友善了不少，但仍有一種情況是始終沒有改變的：就是產房普遍都很冷。產房的室溫真的很冷，只是很多產婦在生產過程之中都會因為痛、因為緊張，往往沒有察覺這點，不會覺得「凍」。等到胎兒娩出後，緊張完了、心定下來了，就會覺得下半身很冷，甚至當上半身也覺得涼時，往往已經着涼或受寒了。胎兒娩出後，產婦會放鬆，痛楚也減少；不過，新媽媽還有很多善後的功夫。這時若不注意保暖，新媽媽就會即時感受到中醫學上說的「風寒」了。

為何產房及手術室總是較為冰冷？

　　這在醫學上是有其需要的，為了保持室內的衛生狀況、對設施之維護、以及藥物的貯存等，室內須維持在一個特定的溫度；而且為了室內空間的清潔乾淨，抽風也要較強，所以為保持空氣的流通，空調風速也相對要調至較強。故此，產婦或病人是不宜要求將室溫調高的。

　　然而，產婦無論在產房生產，或在手術室進行剖腹生產，均是相對衣衫單薄，不會蓋厚被子的，在產房分娩時，下身衣褲會退掉，很可能整個下半身也會暴露出來，而上半身雖然可以穿衣及蓋被，但生產時的用力、痛楚，均會令產婦大量出汗，衣衫自然會濕透。當胎兒娩出後，體內的激素水平會驟然改變，產婦的精神狀況也會放鬆很多，但是，產婦仍需留在產床上完成續後的程序，如：娩出胎盤、處理傷口等。這時如果姑娘不夠細心，陪產者未夠體貼的話，正正就在這個時候會令產婦染上傷風甚至感冒。因為中醫學強調生產後應極之注重保暖、避風，以免感染風寒，引起產後的諸般不適。筆者認為，即使是在

產床上的短短時間，亦應遵從這種產後調護規則，不可稍有鬆懈。很多人以為產後回病房才開始要「包到陷」，這是錯誤的！要保暖、免受風寒，是要在分娩後的一刻開始。有進過產房經驗的人就會知道，躺在產床上等候娩出胎盤及縫合傷口時，是會由腳開始覺得冰冷的，若沒有在這過程中適當地加蓋一些布料為雙腿保暖，雙腿及腳是會受涼的。而已因大量出汗而濕透的上衣，是不能再穿着等候完成傷口處理的，相信讀者都知道在正常情況下汗濕是導致傷風感冒的「幫兇」，何況在產後「百節虛空」、毛孔疏鬆的時刻？因此，陪產者不要只顧着看 BB，也應關心剛努力完的產婦，為她抹汗，如不方便更換上衣，也應用乾毛巾或布，隔開汗濕衣服及皮膚，細心為產婦蓋好被子，特別是下腹部位。生產時因分娩程序的需要，醫生及助產士很可能要求將衣衫及被子拉高以方便工作及保持局部的消毒無菌狀況。但娩出胎兒後，為下腹作適當遮蓋也不礙事。另外，離開產房時，請蓋好被子，手、腳也不應暴露出來，當方便更衣時，應馬上迅速地更換乾爽及保暖的衣褲，並加穿外套，以防被空調「冷親」。

剖腹分娩的保溫之道

手術室的室溫較產房可能更低，因為醫生及護士助手們均要穿上滅菌的手術袍，進行手術時要聚精會神、非常專注，自然需要較清涼的室溫。而剖腹產者又如何可以免受風寒呢？首先，要了解一般的手術程序。當病人躺在手術床上時，護士會為病人調整其體位姿勢以利麻醉師及醫生施行麻醉及手術，很多時候會衣衫不整，也不方便將被子蓋得嚴密。未生產者只要抵抗力不足，就會很易受涼而產後出現傷風感冒症狀，這時如果你覺得冷，可以請姑娘幫忙在肩膊位置加蓋毛巾或毛毯。如果採用「半身麻醉」，在完成麻醉程序後，可請姑娘幫忙加蓋厚毛巾在下身及胸部，而腹部是要留空來進行皮膚消毒及開刀的。若進行「全身麻醉」，請在被「麻醉」之前，告訴護士在手術過程之中為你蓋多些被子。無論用甚麼形式的麻醉，醫生普遍都會用上一些筆者稱之為「肌肉放鬆劑」的藥物，令手術局部的肌肉組織放鬆，以利手術進行。筆者認為使用了這些藥物後，被施藥者的全身腠理疏鬆（即毛孔鬆弛），因此更易受寒着涼。當被麻醉後，人像睡着了一樣，中醫稱之為「邪」的致病物質便會悄悄入侵這個腠理疏鬆的機體；而手術室的室溫是相對偏低的，「寒邪」會更為活躍，所以，一定要在手術中做好保暖工作，免受風寒侵襲。

甦醒後，請不要只顧細味新生兒的種種，更應細心留意自己的手腳是否溫暖，有沒有微涼不溫的感覺，如果有，須立刻請工作人員幫忙加蓋厚被、用暖水袋暖腳，甚至加電毯墊床，總之，就是要為身體加溫保暖，履行中醫學說上的「產後避風防受涼」的古訓。

第二章

本章以中西醫的角度去分析自然分娩與剖腹分娩的區別，對母嬰的好處和壞處，以及全身麻醉和半身麻醉的利弊。採用不同的分娩方式及麻醉方法，產後的護理與調理也有所不同。

自然分娩與剖腹分娩之利弊

何謂自然分娩（陰道分娩）？

　　自然分娩是指經由陰道生產的方式，這是懷胎十月者最期待的一刻。自然分娩也可分為在產程中不需使用器械協助，和在產程中需要醫療器械（如：產鉗、負壓吸引器等）輔助的生產方式。

　　產鉗助產是甚麼呢？當孕婦在娩出胎兒的過程中出現困難時，為了減少危險，醫生有時會用產鉗來輔助，做法是以醫療器械（俗稱產鉗）夾住胎兒的頭部，慢慢牽引、幫助胎兒之頭部離開產道。

　　負壓吸引器又是甚麼？這是一種以負壓吸住胎兒頭部協助生產的辦法。當有需要時，醫生會在助產時以真空吸引器來幫助胎兒娩出。

　　當產程順行時，一般的自然分娩是不需要器械協助的，不過，如果出現胎兒窘迫現象或者是產婦精疲力盡，為免危及產婦或胎兒的生命及健康，此時醫生就必須考慮以器械來協助產婦生產。

妳適合自然分娩嗎？

　　一般沒有非必要原因一定要剖腹的孕婦都可以選擇自然分娩。筆者建議自然分娩應是健康母親的首選生產方式，據統計，自然分娩死亡率遠低於剖腹產，只要你並非患有危疾或因身體缺陷或障礙而不能作自然分娩，在醫學上均可選擇自然分娩。如果因為要替嬰兒擇時辰出生而選擇剖腹產者，筆者建議為了嬰兒的健康着想，切勿選擇離預產期早於四週的日子，這對寶寶出生後的適應力及生長發育都有好處。

自然分娩之優點

　　自然分娩只需局部少量的麻醉或完全不用麻醉藥，產婦經過陣痛及足夠的生理、心理準備之下生產，對產後的恢復較為有利及快速，住院時間較短，後遺症也較少。自然分娩者一般在休息過後就可進食，等到體力恢復後也可以下床，由於身體恢復得快，因此照料嬰兒的精力會較好。自然分娩只是會陰有裂傷，腹痛會較開刀分娩少，因為子宮收縮時只會有宮縮痛，但不會有傷口痛。不用插尿管，因此，尿道及膀胱感染的機會也會較少。因為在生產過程中，有足夠的時間讓身體作出泌乳的準備，對產後的乳汁分泌也有正面作用。

自然分娩對寶寶之好處

自然分娩對嬰兒是有很多好處的，胎兒在母體會接收到子宮收縮陣痛所帶來的訊息，寶寶自己亦會為出生做好準備。陰道分娩特有的子宮收縮和產道擠壓，會令胎兒肺部及呼吸道內的羊水和黏液易於排出，減少了新生兒因吸入羊水、胎糞而引致的呼吸道感染。而且，胎兒胸廓受到子宮收縮所帶來有節律之壓縮和擴張，能促使胎兒肺部產生一種叫做肺泡表面活性物質的東西，使胎兒在出生後的肺泡彈性較好，亦比較容易擴張，因此，自然分娩的嬰兒其呼吸較暢順及肺的功能一般亦比較好。有研究顯示，自然分娩的嬰兒較剖腹分娩的嬰兒，患上呼吸道過敏的機會是較少的。此外，胎兒經過陰道分娩，其頭部必然要受到產道的擠壓，並被拉長變形，這能刺激胎兒的呼吸中樞，令寶寶出生後能立即建立正常呼吸。這種擠壓是為嬰兒由「水中生活」過渡到「陸地生活」所做的準備，隨着嬰兒出生後的第一聲哭叫，肺泡張開，從此開始獨立的呼吸運動，可減少嬰兒因為肺泡表面活性物質缺乏而引起的肺透明膜病變導致新生兒死亡。產道對胎兒頭部的擠壓還可刺激腦活素的釋放，有利於寶寶的智力開發。

自然分娩的嬰兒在離開母體前已為自己的出生作好心理準備，所以自然分娩的嬰兒對外界的燈光、聲音、溫度等的刺激，其反應是較為接受的，因為在未完全離開母體時，他／她們已經習慣了被外界的各種接生行為對待了，故此，較不容易受驚及恐懼。

自然分娩對嬰兒有六大益處

- 分娩過程中子宮的收縮，能讓胎兒肺部得到鍛煉，讓表面活性劑增加，肺泡易於擴張，減少出生後患呼吸系統疾病的機會。
- 子宮收縮及產道的擠壓作用，有助胎兒排出呼吸道內的羊水和黏液，減少新生兒窒息及新生兒肺炎的機率。
- 胎兒頭部經過產道時受到擠壓，令頭部充血可提高腦部呼吸中樞的興奮性，有利於新生兒出生後迅速建立正常呼吸。
- 分娩陣痛使子宮下段變薄，上段變厚，宮口擴張，產後子宮收縮力更強，有利於惡露的排出，也有利於子宮復原。
- 免疫球蛋白 G（IgG）在自然分娩過程中可由母體傳給胎兒，自然分娩的新生兒具有更好的抵抗力。
- 胎兒在產道內受到觸、味、痛覺及本位感的鍛煉，促進大腦及前庭功能發育，對今後運動及性格成長都有好處。

自然分娩之缺點

所有醫療手段都有其優、缺點，自然分娩亦不例外。缺點包括產前的陣痛、生產時下陰撕裂的劇痛等，也許是最令媽媽們害怕的經驗。接生時對會陰的保護不足，有可能令產婦出現肛門或直腸撕裂，以及尿道撕裂等問題。此外，如果生產過程過長的話，亦會增加胎兒窒息的風險。還有可能出現恥骨聯合撕裂、子宮撕裂、脫垂等後遺症，也有產後大出血的可能性。

會陰的裂傷程度可分成四級：第一級裂傷是輕微裂傷，比較少見於初產婦，大多是經產婦；第二級裂傷是中度裂傷，大部分孕婦皆屬於此類的裂傷，一般縫合傷口後問題就不大；第三級裂傷則是指肛門括約肌斷掉，若不修補的話則會造成大便失禁；第四級裂傷則是指肛門也裂開，通常出現在嬰兒過重或有醫療意外等情況下。

何謂剖腹產？為何、何時需要剖腹產呢？

剖腹產俗稱開刀分娩，是自然分娩以外的分娩方式。顧名思義，即是以外科手術在腹部開口，取出胎兒。

在生產的過程中，無論是母體或胎兒任何一方有危險，而經陰道分娩又未能立即成功，即要考慮進行剖腹生產手術。臨床上若預期經陰道生產的過程中，其潛在的危險性較高時，醫生會基於安全考慮建議或指定產婦進行剖腹生產。

需要進行剖腹產的常見原因包括：

- 胎位不正
- 極低體重兒（少於 1500 公克）、巨嬰症
- 特定的胎兒先天畸形（水腦症、腹裂、連體嬰等）
- 多胞胎妊娠
- 孕婦正感染皰疹或其他高傳染性的疾病
- 孕婦嚴重外傷或瀕臨死亡
- 曾經接受過子宮手術
  （上一胎是剖腹生產、子宮肌瘤切除、子宮整形手術等）
- 引產失敗
- 產道阻塞（骨折、腫瘤等）
- 骨盆變形狹窄
- 高齡初產婦（35 歲以上）
- 重度子癇前症或子癇症
- 孕婦罹患心肺疾病等

剖腹產的優點

可按需要的時間來分娩，一切盡在掌控之中，毋須等待「瓜熟蒂落」的時機，亦可避免在自然生產過程中的突發狀況。母體不必經過陣痛，產程較舒服。陰道不受影響，不會有產後鬆弛的問題。對懷巨嬰及雙胞胎的產婦安全較有保障。對一些有腰部舊患及全身性疾患的人較為適合。

剖腹產的缺點

若手術後護理不當，可能的併發症包括：傷口感染、腸黏連及麻醉後遺症。產後身體的恢復較慢，住院時間較長。因為要插尿管，因此增加了尿道感染的機會。因受麻醉的關係，術後不能馬上隨意進食。如施行「插喉」式的全身麻醉，有些人會有咳嗽等問題出現，亦因腹部有傷口，咳嗽時腹痛會加劇。術後亦因腹部有刀口的關係，子宮收縮時會較自然分娩痛。剖腹產因為在子宮上有切口，會有疤痕，所以對於之後的懷孕是會造成不同程度的影響，包括：黏連性胎盤、前置胎盤、胎盤難以剝離等問題之發生率也會增加，以上問題很易引起產前、產後發生出血。因子宮上曾有傷口，在下次的妊娠時，或會有機會出現子宮撕裂、甚至破裂，這對產婦及胎兒來說都是極可怕的情況。因此，一般建議每胎之間最好相隔兩年才懷孕。曾剖腹產者下次未必可以以自然分娩的方式生產，須評估其安全風險才可以決定。

手術的風險

手術的風險包括流血、靜脈炎、傷口感染，若有手術意外，膀胱、直腸、小腸等都有可能受到創傷。麻醉的風險也跟其他外科手術一樣存在。術後當天亦有可能因麻醉反應而出現嘔吐。無論使用哪種方式麻醉，均有一定的危險性，如藥物過敏反應、休克、以及意外等。雖然發生的機會極微，但一旦發生，後果則非同小可，是可以致命的。

全身麻醉

全身麻醉是將麻醉藥物以點滴注射、注射或呼吸吸入等方法為病人施行麻醉，藥物經過血液循環運送到腦部，直接抑制大腦的活動，所以病者在麻醉後是完全處於喪失意識的狀態。麻醉藥物進入循環系統中，所以全身器官（包括腦、心、肺等）都同時受着麻醉藥物的影響。全身麻醉在麻醉師的照顧下一般是很安全的，不過，有些病人在甦醒後會有嘔吐及暫時嗜睡等情況，有些人要等幾天甚至是一週才可完全回復正常。對於剖腹產者來說，若想馬上見到 BB，是不應該作全身麻醉的，但若產婦有下頁列出的情況，則應以安全為上，選擇全身麻醉。

有以下情況者須以全身麻醉為首選：

- 半身麻醉的禁忌症：如血液凝固功能異常（血液不易凝固，易有出血不止的現象）、顱內壓過高（腦部有血塊或腫瘤）等。
- 病人難以配合：採用全身麻醉以外的麻醉方式，病人的意識均是存在的，所以在進行麻醉及手術時病人都必須配合醫生，若病人不合作，則容易產生危險。另外，因為剖腹產的手術可能會要求病人長時間維持在同一姿勢，而其體位也可能不太舒適，在此等情形之下全身麻醉是較適合的。有腰患者，則建議選擇全身麻醉，因俗稱腰麻的麻醉方式是要在腰部打針的，施行麻醉注射時，醫生會要求病人盡量屈曲身體，所以有腰疾者會較為辛苦。

半身麻醉

剖腹產者的另一種常用麻醉方法是半身麻醉，即利用麻醉藥作用在施行手術位置的周邊神經上，暫時性地阻斷神經傳導疼痛至腦部的功能，使病者在手術過程中失去局部感覺。大腦部位完全沒有受到藥物影響，因此，病人是完全清醒的。與全身麻醉相比，局部麻醉具有其獨特的優越性：局部麻醉對神志沒有影響；可起到一定程度的術後鎮痛作用；操作簡便、安全、併發症較少；對患者生理功能影響較小；可阻斷各種不良神經反應，減輕手術創傷所致的刺激反應及恢復較快等。

臨床上醫生一般都會為採用半身麻醉的產婦選擇脊髓麻醉，即是將藥物直接注射至脊髓液中，將神經傳導阻斷，以達止痛效果；以及硬膜外麻醉，即是將藥物注射於脊髓外腔，經滲透進入脊髓液內，其性質與脊髓麻醉近似。這兩種麻醉方式就是一般所説的「半身麻醉」，是目前最為廣泛使用的區域麻醉，主要適用於下肢、下腹部等手術中。而以「半身麻醉」方式進行剖腹產的好處是，寶寶完全不受麻醉藥影響，媽媽因為一直都是清醒的，所以當寶寶準備好時，媽媽便可馬上看見，當寶寶哭出第一聲時，做媽媽的亦可立即聽到，喜悅是全身麻醉者所沒有的。

剖腹分娩的產後調理

剖腹產與自然分娩的最大分別，在於產婦腹部有傷口，因此產後數星期須注意：不宜直接食用薑、忌食生魚和山斑魚、不宜大量進食花膠等。主要原因是有部分孕婦在食用以上食材後，傷口容易長出肉芽，影響傷口癒合及不美觀。

此外，由於在分娩過程中，不論是全身麻醉或是半身麻醉，都使用了麻醉藥。腹部的其他器官（例如：胃、腸、膀肛等）會受到麻醉藥的影響，以致手術後數天未能完全恢復功能。因此產婦在術後首周必需進食較清淡、易消化的食物，一定要確保有肛門排氣（俗稱放屁）及大便。

為避免產婦因便祕而牽動傷口、造成劇痛，剖腹產者應加強攝取含高纖維的食物，喝足夠的水分，維持大腸暢通。

第三章

本章首先介紹 18 道產後補身湯水（P.20-P.55），如沒有特別註明，原則上不論是自然分娩或剖腹分娩的產婦，在十二朝後均可飲用。

進入產後的第二個月，產婦的身體基本上已大致復原了，不過很多餵母乳的媽媽，可能會覺得腰痠背痛或四肢容易疲倦，這時應適當進食多些補肝腎、強筋骨的食療。本章緊接着介紹了 3 道特別適合在產後第二個月飲用的加強版進補湯水。（P.56-P.61）

進入產後第三個月，BB 漸漸長大，喝奶量會增多，母體需要更多的能量去供應寶寶的需要；另外在職的媽媽也要開始重過上班生活，母親的疲累必定會增多。這時媽媽應多補充一些高能量、可增加精神體力的食療湯水（P.62-P.69），以應付繁重的工作。

即時睇片

木瓜牛奶鯇魚濃湯

催乳補血
十二朝前
剖腹產者去薑即可

功效：助分泌乳汁，輕補而不燥熱，可補血助產後復原。

服法：天天喝也無妨。

宜忌：不想哺乳者，不宜多喝。

材料

本地半熟木瓜 2 斤

脫脂牛奶 800 毫升

鯇魚尾 1 條（1.5-2 斤）

生薑 3 片

陳皮一小塊

紅棗 5 粒（去核）

生油少許

水 2000 毫升

做法

1. 木瓜去皮、去籽，切件。
2. 鯇魚尾去鱗、洗淨，用少許油煎至兩面微黃即取出，用廚紙將魚的油分抹去，備用。
3. 陳皮浸軟，刮去白瓤。
4. 注清水入高身湯煲內，加入木瓜、鯇魚尾、生薑、陳皮及紅棗，以大火煮滾後，改中慢火煮 30 分鐘，加入脫脂牛奶再煮 20 分鐘，下鹽調味即成。

椰子紅豆通草鮮八爪魚湯

功效：催乳、滋補，助恢復體力。

服法：隔天一次。

宜忌：不想哺乳者，不宜多喝。

材料

老椰子肉 1 斤

鮮八爪魚 3 斤

紅豆 4 両

通草 2 両

瘦肉半斤

生薑 3 片

陳皮一小塊

紅棗 5 粒（去核）

水 2000 毫升

做法

1. 鮮八爪魚洗淨、去淨內臟，用粗鹽洗擦去潺，切件，汆水，備用。

2. 瘦肉洗淨、汆水，切件備用。

3. 陳皮浸軟，刮去白瓤；藥材用煎藥用紙袋或煲魚湯用紗袋包好。

4. 注清水入高身湯煲內，加入全部材料，以大火煮滾後，改中慢火煮 60 分鐘，下鹽調味即成。

通草
母親產後經絡不調，以致氣血虛弱、乳汁不足。通草具
有清熱利尿、通氣下乳的功效。

冬蟲夏草烏雞燉湯

全面補身
滿月後

功效：固本培元，補脾胃、固肺腎。有助改善虛弱體質。
服法：每週 1-2 次。
宜忌：外感者不宜。

材料

冬蟲夏草 5 錢
南棗 4 粒
淮山 1 両
烏雞半隻
生薑 2 片
凍開水 1000 毫升

做法

1. 烏雞洗淨、去皮、去內臟，剪掉雞翼尖、頭頸及臀部，
   洗淨，汆水，瀝乾水分，切中件備用。
2. 冬蟲夏草、南棗、淮山洗淨。
3. 所有材料放入燉盅內，蓋好蓋，隔水以文火燉 4 小時，
   下鹽調味即成。

冬蟲草

產婦在坐月期間體質較虛，冬蟲草是
極佳的調補藥材。產婦容易出現筋骨
疲痛、渾身乏力的現象，冬蟲草有助
補氣血、增強體質，提昇免疫力。

杞子桂圓桑寄生蓮藕湯

補肝腎、壯筋骨
十二朝後

功效：補肝腎、壯筋骨。養血明目，補血。有助改善及預防產
後腰痛。

服法：每週 1-2 次。

宜忌：外感者不宜。

材料

杞子半両

桂圓肉 4 錢

桑寄生 1 両

有機蜜棗 4 粒

蓮藕 1.5 斤

西施骨 2 斤

生薑 2 片

水 3000 毫升

做法

1. 西施骨洗淨，汆水，去淨油脂。
2. 蓮藕去皮、洗淨，切中件備用。
3. 杞子、桂圓肉、桑寄生洗淨備用；藥材用煎藥用紙袋或煲魚
   湯用紗袋包好。
4. 注清水入高身湯煲內，加入全部材料以大火煮滾後，改中慢
   火煮 90 分鐘，下鹽調味即成。

桂圓肉

味甘性溫，能補益心脾、養血安神，主治失眠健忘、病後及產後血氣不足等，也可治療因心血虛引起的神經衰弱症。

北芪黨參海玉竹花膠湯

花膠

花膠含豐富蛋白質及膠質，滋補
療效高，產婦多吃能滋補身體，
令傷口迅速復原。

功效：補氣固表，滋陰養顏。此湯應多喝，可令肌膚保持嬌嫩
潤澤；有延緩衰老、維持肌膚彈性、益脾、助產後復原
的作用。

服法：每週 1-2 次。

宜忌：外感者不宜。

材料

炙北芪 4 錢
黨參 5 錢
急凍海玉竹頭 4 両
乾花膠 2 両
瘦肉半斤
有機蜜棗 4 粒
陳皮一小塊
生薑 2 片
水 3000 毫升

浸發花膠用料

薑 6 片
葱 2 棵
紹酒半碗
清水 2 大煲

做法

1. 乾花膠浸水過夜，換水再浸 8 小時，剪開去淨內裏污物。
   燒滾一鍋清水，放入花膠、熄火，焗水至水凍，取出花
   膠剪成數份。再燒滾一鍋清水加薑及葱同煮，放入花膠
   滾 3 分鐘，灒酒、滾 2 分鐘，熄火、焗水至凍，取出沖
   洗花膠至乾淨為止。
2. 海玉竹洗淨、去皮、切片備用。瘦肉汆水，切件。
3. 用清水浸北芪、黨參 2 小時，水棄掉不要。
4. 陳皮浸軟，刮去白瓤。
5. 全部材料放煲內加清水以大火煮滾，改慢火煮 3 小時，
   下鹽調味即成。

石斛女貞子天冬北沙參南芪豬腱湯

滋陰潤補
十二朝前
剖腹產者去薑改用陳皮即可

功效：滋陰潤燥，可改善口乾、便秘、虛火等問題。
服法：每週 2-3 次。
宜忌：外感者不宜。

材料

藿山石斛 4 錢
女貞子 5 錢
天冬 1 両
北沙參 1 両
南芪 1 両
豬腱 1 斤
有機蜜棗 4 粒
生薑 2 片
水 2000 毫升

做法

1. 豬腱洗淨，汆水，再洗淨瀝水備用。
2. 藿山石斛、女貞子、天冬、北沙參、南芪洗淨，瀝乾水；天冬、北沙參加水浸 2 小時，水棄掉不要。藥材用煎藥用紙袋或煲魚湯用紗袋包好。
3. 全部材料加水以大火煮滾，改文火煮 2 小時，下鹽調味即成。

女貞子

味甘帶微苦，性涼，有滋補肝腎、清熱明目的功效。女貞子是補益類中藥，適宜女性產前產後調養之用。尤其是產後掉髮、白髮增多者，也對產後耳鳴有幫助。

田七熟地紅棗百合牛腱湯 補血活血｜十二朝後

熟地

熟地味甘、性微溫，能補血滋陰、補精益髓、治陰虛血少，對於子宮的復原有很顯著的功效。

功效：補血活血。有助子宮復舊，改善惡露排出不暢的問題。補而不燥，活血而不會太行血。

服法：喝湯吃牛腱。牛腱可切片拌淡豉油調味食用。每週 1-2 次。

宜忌：血性惡露過多者不宜。

材料

熟田七 3 錢

熟地 5 錢

紅棗 10 粒（去核）

百合 1 両

牛腱 1 斤

陳皮一小塊

有機蜜棗 4 粒

生薑 2 片

水 2500 毫升

做法

1. 牛腱洗淨，燒滾水後、熄火、下牛腱，燙 1 分鐘，取出再洗淨備用。
2. 熟田七、熟地、紅棗洗淨；百合浸水 2 小時，水棄掉不要；藥材用煎藥用紙袋或煲魚湯用紗袋包好。
3. 全部材料加水以大火煮滾，改文火煮 2 小時，下鹽調味即成。

杜仲肇實田雞湯 除煩健腎
十二朝後

功效：除煩健腎，溫和活血化瘀。可改善心情，亦有助去瘀生新，健腎益精。

服法：每週 1-2 次。

宜忌：外感者不宜。

材料

有機乾焙金針 4 錢
乾木耳 1 兩
鹽水製杜仲 4 錢
鹽水製肇實 1 兩
田雞半斤
生薑 3 片
紅棗 5 粒（去核）
糯米酒 2 湯匙
水 1000 毫升

做法

1. 田雞去皮、去脊骨、去爪，洗淨、汆水，瀝乾水備用。
2. 燒紅油鑊，下薑爆香後加入田雞炒 2-3 分鐘，以去泥腥味。
3. 乾木耳浸軟、去硬蒂，洗淨，撕成適口大小。金針洗淨。
4. 藥材用煎藥用紙袋或煲魚湯用紗袋包好。
5. 全部材料加水以大火煮滾，改文火煮 20 分鐘，加酒續煮 20 分鐘，下鹽調味即成。

杜仲

杜仲味甘、性溫，有補肝腎、壯筋骨的功效，有助產婦活血化瘀。

冬菇海參鮑魚湯

滋陰平補
十二朝前
剖腹產者去薑即可

功效：滋陰補陰，養血益氣。特別適合陰虛火旺者用來補身。

服法：天天喝也無妨。

宜忌：外感者不宜。

貼士：海參可選用乾貨，但要浸發透才可煮湯。

材料

優質乾冬菇 10 個

急凍海參約 1.5 斤

澳洲急凍鮑魚約 1 斤

瘦肉半斤

花生肉 4 両（連衣）

陳皮一小塊

生薑 3 片

水 2500 毫升

做法

1. 所有材料洗淨，冬菇浸軟、洗淨，分開菇身及硬蒂，菇身厚者，可剪開，以利出味。瘦肉汆水、去油脂，切厚件。鮑魚去內臟、洗淨，切厚片。

2. 陳皮浸軟，刮去白瓤。

3. 急凍海參解凍後，剪開、去淨內臟及污物，洗淨、加紹酒汆水 2 分鐘。

4. 全部材料加水以大火煮滾，改文火煮 2 小時，下鹽調味即成。

海參

海參有補腎益精、滋陰潤燥的功效，又具有增血作用，海參所含的膠質對補充體力具有良效，十分適宜產婦食用。

黑豆川芎天麻鯛魚湯 祛頭風、治頭痛
十二朝後

功效：祛頭風、治頭痛，輕微行血、化瘀，有助改善頭重、頭脹等問題。
服法：隔天一次，直至問題改善。
宜忌：血性惡露過多者慎服。

材料

黑豆 2 両
乾野生小雲耳 2 両
川芎 3 錢
大棗 4 粒
薑汁製天麻 4 錢
白芷 3 錢
鯛魚 2 斤
生薑 4 片
水 2300 毫升

做法

1. 鯛魚去鱗、去內臟及鰓，洗淨、抹乾，用少許油煎魚至兩面微黃即取出，用廚紙將魚的油分抹去，備用。
2. 小雲耳浸軟、去硬蒂，洗淨。
3. 藥材用煎藥用紙袋或煲魚湯用紗袋包好。
4. 全部材料加水以大火煲滾後，改文火煮 2 小時，下鹽調味即成。

天麻

天麻味辛、性溫，有祛風、去濕、止痛、行氣活血的功效，更能鎮靜神經，有助產婦調理身心。

蓮子紅豆通草三文魚頭湯

催乳補虛
十二朝前
剖腹產者去薑改用陳皮即可

功效：日常催乳湯，可保持乳汁分泌，亦有補虛的功效。

服法：天天喝也無妨。

宜忌：外感者去雞不用。

材料

蓮子肉、紅豆 2 両（連衣）
三文魚頭 1 斤
雞 1 隻約 1.5 斤
通草 3 錢
生薑 3 片
水 2200 毫升

做法

1. 三文魚頭洗淨切件，用少許油煎香，抹去油分備用。
2. 雞洗淨、去皮、去內臟，剪掉雞翼尖、頭頸及臀部，洗淨、汆水，去油脂備用。
3. 蓮子肉洗淨備用。
4. 全部材料加水以大火煲滾後，改文火煮 2 小時，下鹽調味即成。

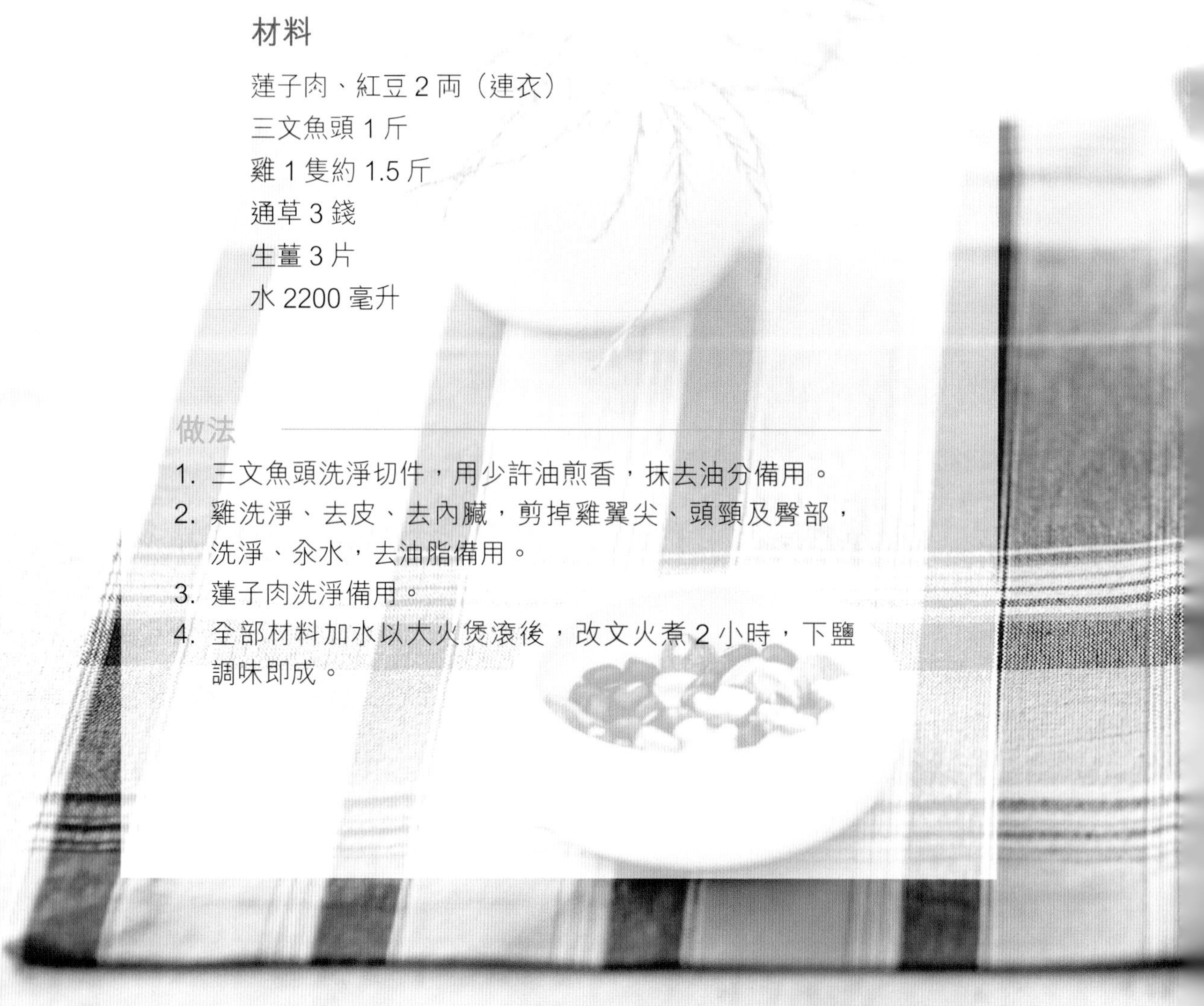

黃鱔花生湯

補腎益血
十二朝前
剖腹產者去薑改用陳皮即可

功效：有助乳汁分泌，補腎益血。養血補體力，可作日常催乳之用。

服法：每週 1-2 次。

宜忌：外感者不宜。

材料

黃鱔 1 斤

花生 4 両（連衣）

紅棗 5 粒（去核）

王不留行子 3 錢

瘦肉 4 両

水 2200 毫升

生薑 3 片

調味料

米酒 1 湯匙

胡椒粉 1/2 茶匙

鹽 1/3 茶匙

做法

1. 瘦肉洗淨、汆水，切件備用。

2. 黃鱔去內臟，洗淨，用粗鹽搓擦外皮及內部去潺；再洗淨，切件備用。燒紅油鑊爆香薑片及黃鱔，用廚紙將魚的油分抹去，備用。

3. 花生、紅棗、王不留行子洗淨備用；用煎藥用紙袋將王不留行子包好。

4. 全部材料加水以大火煲滾後，改文火煮 1 小時，下調味料拌勻即成。

王不留行子

王不留行子是草本植物麥藍菜果實的乾燥成熟種子，味苦、性平。《本草綱目》指出其名字的由來是因為藥性走而不守。能通利血脈和乳汁，故有催乳之效。

核桃雪耳無花果響螺湯

潤補肝腎
十二朝前
剖腹產者去薑即可

功效：潤補肝腎，增氣力，亦有助保障泌乳充足。

服法：天天喝也無妨。

宜忌：外感者不宜。

材料

核桃 2 両（連衣）

乾雪耳 2 両

有機生曬無花果 6 粒

澳洲急凍響螺肉 500 克

鮮雞腳 8 隻（中）

陳皮一小塊或生薑 4 片

水 2800 毫升

做法

1. 雪耳浸軟，去硬蒂及雜質，撕成數塊。
2. 無花果浸軟，剪成兩份。
3. 急凍響螺肉解凍、汆水，切件備用。
4. 雞腳洗淨、汆水，再洗淨瀝水備用。
5. 陳皮浸軟，刮去白瓤。
6. 全部材料加水以大火煲滾後，改文火煮 1.5 小時，下鹽調味即成。

番茄雞肉大豆芽山楂湯

開胃通便
十二朝前
剖腹產者亦宜

功效：開胃增進食欲，纖維素豐富，有助保持大便暢通。
服法：天天喝也無妨。
宜忌：外感者以豬肉代替雞肉。

材料

番茄約 10 兩
雞胸肉半斤
大豆芽菜 4 兩
山楂肉 5 錢
韭葱 2 兩
水 900 毫升
鹽少許
砂糖 1 茶匙

做法

1. 雞胸肉洗淨、去筋膜，切件備用。
2. 將其餘材料洗淨備用。番茄切中件備用。用煎藥用紙袋將山楂包好。
3. 韭葱及大豆芽菜去根部；韭葱切細絲。
4. 山楂、番茄及雞胸肉加水以大火煲滾，改中火煮 20 分鐘，下大豆芽菜煮至熟，下韭葱、鹽及糖煮片刻即成。

金銀百合南瓜素湯

潤肺補腎
十二朝前
剖腹產者去薑改用陳皮即可
全素食

功效：適合素食者的補身湯水。潤肺、補腎，有飽肚感。

服法：天天喝也無妨。

材料

鮮百合 4 個

乾百合 1 兩

栗子肉 2 兩

黑豆 4 兩

南瓜肉 1.5 斤（切件）

生薑 2 片

水 2000 毫升

做法

1. 乾百合、栗子肉及黑豆浸水約 1 小時，水倒掉不要。
2. 鮮百合、南瓜肉洗淨備用。
3. 全部材料（鮮百合除外）加水以大火煲滾後，改文火煮 1 小時，下鮮百合煮 10 分鐘，下鹽調味即成。

乾榆耳羊藿葉素湯

功效：適合素食者的補身湯水。壯腎、益精、補血。

服法：天天喝也無妨。

宜忌：內熱盛者不宜多喝。

材料

乾榆耳 2 両

乾冬菇 1 両

紅菜頭 1 斤

杞子 5 錢

淮山 2 両

椰子肉半斤

羊藿葉 4 錢

生薑 3 片

水 2300 毫升

做法

1. 全部材料洗淨。乾榆耳及乾冬菇浸軟，洗淨，將菇身及硬蒂分開；榆耳去硬蒂。
2. 紅菜頭去皮、洗淨，切件。椰子肉切件。
3. 用煎藥用紙袋將羊藿葉包好。
4. 全部材料加水以大火煮滾後，改文火煮 1.5 小時，下鹽調味即成。

乾榆耳 ● 羊藿葉

乾榆耳有健脾養顏、滋陰補腎的功效，
含有幫助吸收天然食物之膠質，有助
產婦復原。羊藿葉能助陽益腎、祛風
除濕，增加荷爾蒙分泌，有助產婦舒
緩乳汁不通、目昏、水腫等症狀。

鹿茸人參燉老雞

大補元氣
滿月後

功效：大補元氣，氣血相補。

服法：每週一次。

宜忌：內熱盛者及外感者不宜。

材料

鹿茸 1 錢

人參 1.5 錢

老雞半隻

老薑 2 片

凍開水 3 碗

做法

1. 老雞洗淨、去皮、去內臟，剪掉雞翼尖、頭頸及臀部，洗淨，汆水，瀝乾水分，切中件備用。
2. 所有材料放入燉盅內，蓋好蓋，隔水以文火燉 4 小時，下鹽調味即成。

人參

人參有大補的功效，可助產婦補血補氣，然而，人參會減少乳汁的分泌，欲餵哺母乳者，不建議服用。

珍珠肉石斛海參湯

滋胃補腎
十二朝前
剖腹產者去薑改用陳皮即可

功效：滋胃陰，補腎陰，解口渴。有助傷口復原。

服法：天天喝也無妨。

宜忌：外感者不宜。

材料

珍珠肉 1 両
藿山石斛 3 錢
急凍海參半斤
瘦肉 1 斤
生薑 3 片

做法

1. 瘦肉汆水、去油脂，切件。珍珠肉浸水半小時，洗淨，對半剪開。
2. 急凍海參解凍後，剪開、去淨內臟及污物，洗淨、加紹酒汆水 2 分鐘。
3. 全部材料加水以大火煮滾，改文火煮 2 小時，下鹽調味即成。

石斛

石斛有養陰清熱生津作用，因為品種、產地和加工方法的不同而有不同名稱。產於安徽霍山的稱為「霍山石斛」，藥性較平和，滋陰生津力較好，尤適合不耐寒者使用。

杜仲牛大力金狗脊湯 壯筋骨、去瘀痛

功效：壯筋骨，去痠痛，減少媽媽手及腰背不適，為媽媽復工
作準備。

服法：可每週喝 1-2 次。

材料

厚杜仲 5 錢

牛大力 4 錢

金狗脊 3 錢

蜜棗 6 粒

陳皮一小塊

豬尾骨 1-1.5 斤

水 2500 毫升

做法

1. 藥材用煎藥用紙袋或煲魚湯用紗袋包好。
2. 陳皮浸軟，刮去白瓤。
3. 豬尾骨洗淨、汆水、去油脂，切件。
4. 全部材料加水煮滾後改文火煮 2 小時，下鹽調味即成。

黑色五珍湯

烏髮養顏
剖腹產者去薑即可

功效：補肝腎，烏髮、減少掉髮。很多媽媽產後第 2 個月開始
　　　出現掉髮增多現象，此時多喝這款湯水，有促進頭髮生
　　　長的功效；亦可養顏，改善面色，減少容顏憔悴。
服法：可每週喝 1-2 次。

材料

製何首烏 3 錢
黑豆 2 両
黑木耳 1 両
黑棗 4 粒
烏雞一隻
生薑 3 片
水 2200 毫升

做法

1. 黑木耳洗淨、去硬蒂，撕成適口大小。
2. 烏雞去皮、去內臟、去頭頸、翼尖及臀部，洗淨、汆水。
3. 全部材料加水煮滾後改文火煮 2 小時，下鹽調味即成。

高麗參鹿茸燉瘦肉

固本培元
剖腹產者去薑即可

功效：具大補氣血，固本培元的功效。適合體虛媽媽在產後 2
個月開始，每週大補一次，為湊日益胖壯、喝奶量日多
的寶寶作好準備，鞏固好母親的身體，增強體質，減少
日後患病的機會。

服法：每週喝一次。

材料

高麗參 2 錢
鹿茸 1 錢
陳皮一小塊
生薑 2 片
紅棗 4 粒（去核）
瘦肉半斤
開水 1000 毫升

做法

1. 陳皮浸軟，刮去白瓤。
2. 瘦肉洗淨、汆水、切件。
3. 全部材料加水入燉盅封蓋，隔水以中火燉 4 小時，取出
   下鹽調味即成。

八珍鮑魚海魚湯

滋陰養陽
剖腹產者去薑即可

即時睇片

功效：氣血雙補，既滋陰又養陽，是媽媽在產後 100 天內的最佳補身方法。

服法：產後 3 個月開始每 1-2 週服一次。

材料

當歸、熟地、雲苓、黨參各 3 錢

川芎、白芍、白朮各 2 錢

甘草 1.5 錢

大棗 5 粒

鮑魚 1-2 隻

任何有鱗的海魚 1 斤

生薑 3 片

水 2800 毫升

做法

1. 魚去鱗、洗淨，用少許油煎至兩面微黃即取出，用廚紙將魚的油分抹去，備用。
2. 鮑魚去內臟、洗淨，切件，備用。
3. 藥材用煎藥用紙袋或煲魚湯用紗袋包好。
4. 全部材料加水以大火煮滾後，改文火煮 2 小時，下鹽調味即成。

當歸 ● 黨參

當歸具良好的補血、活血及滋補子宮之效，是產婦常用的補身食材，但不宜於坐月早期（十二朝前）進補，建議順產者於惡露乾淨後進食；剖腹產者則一個月後食用。黨參味甘性平，能補脾益氣、生津養血，並可增強抵抗力，能改善手腳冰冷，令面色紅潤。

十全大補雞湯

溫經祛寒
產後三個月後始服用，剖腹產者亦宜

川芎
味辛性溫，有補氣通脈、行氣活血、
祛風止痛的功效，有助產婦復原。

功效：大補元氣，溫經祛寒的補湯，是媽媽在產後 100 天內的
防受寒、祛風寒的必然選擇。特別適合平素怕冷、畏寒
的女士，另外，在冬天坐月的媽媽也可略為多喝此湯。

服法：產後 3 個月開始每 1-2 週服一次。

材料

高麗參、當歸頭各 2 錢

肉桂 1 錢

熟地、雲苓各 3 錢

川芎、白芍、北芪、白朮各 2 錢

甘草 1.5 錢

大棗 5 粒

老母雞一隻

生薑 3 片

陳皮一小塊

水 3000 毫升

做法

1. 陳皮浸軟，刮去白瓤。
2. 雞去皮、去內臟、去頭頸、翼尖及臀部，洗淨、汆水。
3. 藥材用煎藥用紙袋或煲魚湯用紗袋包好。
4. 全部材料加水以大火煮滾後，改文火煮 2 小時，下鹽調
味即成。

龜鹿牛骨湯

滋腎壯骨
剖腹產者去薑即可

功效：滋腎，壯骨，強健筋腱，補腳力。

服法：產後 2-3 個月開始，母體的筋骨一般需要加強保護，多喝此湯有助預防筋骨關節不靈活等小毛病。可每週喝一次，記得把鹿筋也吃掉，補筋骨的效果才更明顯。

材料

陳龜板一塊（大）
乾鹿腳筋約 6 両
牛骨 2 斤
紅棗 8 粒（去核）
陳皮一小塊
生薑 3 片
瘦肉 1 斤
水 3000 毫升

做法

1. 陳皮浸軟，刮去白瓤。
2. 鹿腳筋用清水浸軟、洗淨、去蹄甲及雜質，汆水一次；取出撕去筋膜，剪去腐肉及血管等，洗淨；加入薑 3 片或葱 2 棵再汆水片刻，撈起，趁熱潷酒半碗拌勻，用清水沖洗淨，以去除腥味，切件，備用。
3. 牛骨洗淨、汆水、去油脂。
4. 瘦肉洗淨、汆水、切件。
5. 龜板洗淨。
6. 全部材料加水以大火煮滾後，改文火煮 2.5 小時，下鹽調味即成。

龜板

龜板味甘鹹、性微寒，主要功效之一是補腎健骨，用於腰膝無力、筋骨痿軟等。

補腎壯筋骨素湯

補腎壯筋
十二朝後至 100 日
全素食

功效：適合素食者的補身湯水，具有補腎、壯筋骨的功效。

服法：每週 2-3 次。

宜忌：外感者不宜。

材料

杜仲 3 錢

黑豆 1 両

合桃 1 両

栗子肉 2 両

紅棗 10 粒（去核）

蜜棗 5 粒

乾雪耳 1 両

陳皮一小塊

水 2500 毫升

做法

1. 雪耳浸軟，洗淨、去硬蒂，撕成適口大小。
2. 陳皮浸軟，刮去白瓤。
3. 全部材料加水以大火煲滾後，改文火煮 1.5 小時，下鹽調味即成。

第四章

除了補健湯水，食療對於促進產婦的復原也是十分重要的。尤其是初為人母的產婦或陪月員，經常為產後的膳食而苦惱：哪些食材可以進食？一日三餐的菜式可否有多些變化？哪些食物有助母乳分泌？本章介紹的養生食譜，正好為大家解決疑難。

八爪魚杞子雞蛋飯

催奶補血
十二朝前
剖腹產者用薑汁醃八爪魚辟腥後沖走，
不加薑絲一同焗飯即可食用

功效：助上奶，補血而不燥熱，增加產婦體力。亦極適合作五更飯食用。冬夏皆宜。

服法：天天吃也無妨。可作五更飯食用。

宜忌：外感者不宜。

材料

鮮八爪魚半斤
杞子 1 湯匙
雞蛋 1 隻
珍珠米 1/3 杯
薑絲 1 湯匙
黑、白芝麻約各 1 茶匙
生油 1/2 茶匙
蔥花少許

醃料

砂糖、鹽各 1/3 茶匙
麻油、豉油各 1/2 茶匙
米酒 1 湯匙

做法

1. 八爪魚洗淨、去淨內臟，用粗鹽洗擦去潺、汆水、切粒，加醃料醃十分鐘。燒熱油鑊，下薑絲爆香，下八爪魚粒炒香至半熟；離鑊備用。

2. 米洗淨，放入電飯煲內煮至水收乾，加入炒過的八爪魚粒、杞子及雞蛋（打勻）同煮至飯熟、電飯煲跳掣。

3. 用白鑊炒香黑、白芝麻。

4. 在飯上撒上蔥花及已炒香的黑、白芝麻即可享用。

蒜茸三文魚扒意粉

補鈣暖胃
十二朝前
剖腹產者亦宜

功效：補鈣暖胃，適合所有產婦服用。剖腹產者在醫生允許進食正常食物後即可食用這款食譜。

服法：天天吃也無妨。

宜忌：對食材過敏者不宜。

材料

有機三文魚扒一件（約 300 克）
蒜茸 1 湯匙
乾意大利粉適量（按個人胃口而定）
青葱 1 棵（切成葱花）
水約 1000-2000 毫升
（跟隨意粉包裝上指示）
鹽 1/3 茶匙（煮意粉時用）
生油 1/2 茶匙

醃料

檸檬汁 2 茶匙
麻油 1 茶匙
鹽 1/3 茶匙
黑胡椒粉適量

做法

1. 將三文魚扒洗淨、抹乾備用。
2. 用醃料將三文魚肉醃十分鐘。
3. 按包裝指示將意大利粉煮至個人喜歡的口感熟度，瀝水備用。
4. 燒紅油鑊，用大火煎三文魚扒 1 分鐘，反面再煎 1 分鐘，取出備用；原鑊加入蒜茸爆香；改中火、下已煮熟的意粉快速兜勻，加入三文魚扒略煎片刻至全熟，撒上葱花即可上碟。

陳皮免治牛肉紅棗飯

紅棗
性平，有補血及恢復精力的功效，產
後喝紅棗水有助補身，建議先去核可
減低燥熱。

功效：補血，增加體力。

服法：每週 1-2 次。

宜忌：熱氣者慎服。

貼士：若不喜歡吃到陳皮，可於進食前把陳皮絲棄掉。

材料

陳皮絲 1/2 湯匙
免治牛肉 200-300 克
紅棗約 10 粒（去核）
白米 1/3 杯
葱花 1-2 茶匙

醃料

白胡椒粉、砂糖、
鹽各 1/3 茶匙
麻油、豉油各 1/2 茶匙
米酒 1 湯匙
水 1/2 湯匙

做法

1. 將免治牛肉加入已拌勻的醃料醃 30 分鐘。

2. 紅棗切細粒；陳皮浸軟，刮去白瓤、切絲。把紅棗粒、陳皮絲與牛肉拌勻，一起醃 30 分鐘。

3. 米洗淨，放入電飯煲內煮，當米收乾水（約 8 成熟左右），將牛肉放入飯面焗至飯熟，電飯煲跳掣，在飯上撒上葱花即成。

薑汁黃鱔薑黃焗飯

暖身祛寒
自然分娩者，十二朝前可食用
剖腹產者，十二朝後才可食用

功效：益精養血，暖身止嘔，祛寒補體力。

服法：每週 1-2 次。

宜忌：因薑及薑汁的用量較多，怕辣或很易上火者可按個人體質減少
薑的用量，但不能完全不用。酒精過敏者將糯米酒的量減至 1
湯匙。

材料

薑汁 1/3 碗
黃鱔半斤
糯米酒 3 湯匙
白米 1/3 杯
薑絲 1/2 湯匙
生油 1/2 茶匙

醃料

砂糖、鹽各 1/3 茶匙
麻油、豉油各 1/2 茶匙
薑黃粉 1/2 茶匙

做法

1. 黃鱔剖開，洗淨，用粗鹽擦淨內外去潺；再洗淨，去骨，再
   洗淨骨碎；切成 2 吋長段，加入已拌勻的醃料醃 30 分鐘。
2. 米洗淨放入電飯煲內煮。
3. 燒熱油鑊、下薑絲爆香，下醃好的黃鱔爆炒至五成熟，下薑
   汁，改文火瀡酒、煮至酒味揮發掉，取出離鑊備用。
4. 待電飯煲內的飯收乾水時，將炒過的黃鱔放入飯面繼續煮至
   飯熟，電飯煲跳掣即成。

松子菜粒肉碎炒飯

性質溫和
十二朝前
剖腹產者去薑，改用乾葱起鑊即可

功效：材料性質溫和，適合任何體質者服用。味道清淡易被接
　　　受，常服也沒有任何不妥。

服法：天天吃也無妨。

宜忌：對松子過敏者不宜，可以其他果仁代之。

材料

松子仁 1 湯匙
菜芯梗約 10 棵
免治豬肉 200 克
白飯 1-1.5 碗
葱花 2 茶匙
葡萄核油 1 茶匙
生薑 3 片

醃料

白胡椒粉、砂糖、
鹽各 1/3 茶匙
麻油、豉油各 1/2 茶匙
白酒 1 茶匙
水 2 茶匙

做法

1. 松子用白鑊炒香。
2. 菜芯梗洗淨、切粒備用。
3. 用手搯鬆白飯，把飯粒弄散。
4. 拌勻醃料後，加入瘦肉中拌勻，醃 30 分鐘；燒熱油鑊，
　　炒肉碎至八成熟，離鑊備用。
5. 燒熱油鑊，爆香薑茸，下白飯用大火兜炒 3 分鐘，收中
　　火、下已炒過的肉碎、菜粒，炒至肉、菜全熟，撒葱花
　　炒勻即可上碟，食前撒上炒香的松子。

栗子冬菇田雞腿紅米飯

補腎行血
十二朝前
剖腹產者去薑改用陳皮即可

功效：補腎，輕微行血，增加膳食纖維，營養豐富，補充維他命 B 雜。

服法：天天吃也無妨。宜作五更飯食用。

宜忌：對食材過敏者不宜。

材料

栗子肉 10 粒
冬菇 4 個
田雞腿 5 對
乾雲耳約 3-5 朵
紅米 1/3 杯
薑絲 1 湯匙

醃料

白胡椒粉、砂糖、
鹽各 1/3 茶匙
麻油、豉油各 1/2 茶匙
白酒 1 茶匙

做法

1. 田雞腿洗淨，抹乾，拌入已拌勻的醃料醃 30 分鐘。

2. 乾雲耳浸軟，去硬蒂，洗淨，撕成適口大小。冬菇浸軟，洗淨，去硬蒂後切絲。栗子肉去衣、洗淨，壓碎成粗粒。

3. 米洗淨放入電飯煲內、加入栗子粒、冬菇、雲耳及薑絲同煮，當飯收乾水時（約 7 成熟左右），將田雞腿放入飯面焗至飯熟，電飯煲跳掣即成。

香蒜排骨冬菜蒸糙米飯

纖維豐富
十二朝前
剖腹產者亦宜

功效：鮮香好味，膳食纖維豐富，能保持大便暢通。
服法：天天吃也無妨，是理想的五更飯選擇之一。
宜忌：消化力差者，糙米要浸水久些，以免飯粒過硬。

材料

蒜頭 2-3 瓣
排骨約 250 克
冬菜 1 茶匙
糙米 1/3 杯

醃料

白胡椒粉、砂糖各 1/3 茶匙
鹽 1/4 茶匙
生粉、豉油、麻油各 1/2 茶匙
米酒 1 茶匙

做法

1. 蒜頭拍扁、去衣，剁成蒜茸。
2. 排骨洗淨，切成適口大小；加入已拌勻的醃料、蒜茸及冬菜醃 60 分鐘。
3. 糙米洗淨、浸水 2 小時；放入電飯煲內煮，待煮至剛收水時，加入醃好的排骨，繼續煮至飯熟、電飯煲跳掣，焗 5 分鐘，再次按掣煮至再跳掣即成。

健康臘味生炒糯米飯

補中益氣
十二朝前
剖腹產者亦宜

功效：補中益氣，健康美味。補身之餘，每週吃 1 至 2 次也不易增磅。糯米有補身健體的作用。

服法：每週 1-2 次。

宜忌：對花生過敏者可以不下花生，改用炒香的欖仁，味道更佳。

貼士：選用加瘦臘腸及瘦叉燒，因本身已有味道，所以不用再下太多的調味料，味道也不會單調。

材料

加瘦臘腸 1 條
瘦叉燒 100 克
冬菇 1-2 個
蒜子肉 2 瓣
芥花籽油 1 茶匙
糯米 1/3 杯
水 1/2 杯
葱 2 棵
炒香花生約 10 粒

調味料

老抽、砂糖各 1/4 茶匙
生粉、豉油、麻油各 1/3 茶匙
白酒 1 茶匙

做法

1. 糯米浸水 4 小時。冬菇浸軟後，去硬蒂，切粒。把臘腸放入熱水中略浸洗、去油、切粒。瘦叉燒切粒。

2. 蒜頭拍扁、去衣，剁成蒜茸。

3. 下少許油在易潔鑊中，爆香蒜茸；下臘腸粒炒熱；下冬菇炒熟，盛起備用。

4. 原鑊，改中火下已瀝乾水分的糯米，快速兜勻，盡量鋪平糯米在鑊中，讓糯米平均受熱。分多次加入清水，每次約 1-2 湯匙。然後炒勻之。重覆加水炒至糯米熟透（約需 25-35 分鐘）；最後加調味料兜勻、炒至汁乾，下已炒過的材料及叉燒粒再兜勻，撒上葱花及花生粒拌勻，即可上碟。

意大利火腿燜津白

有益腸道
十二朝前
剖腹產者只用薑起鑊，下菜前取走即可

功效：美味、有營。纖維素豐富，有助腸道健康。

服法：天天吃也無妨。

貼士：若火腿帶肥，炒菜的油可酌減。也可用西班牙火腿代替意大利火腿。這個菜傳統上是用雲南的宣威火腿或金華火腿入饌，但要確定是優質安全食材才採用。

材料

意大利火腿約 50 克

黃芽白（紹菜）300 克

核桃油 1 茶匙

鹽 1/4 茶匙

蒜子肉 3 瓣

生薑 3 片

粟粉 2 茶匙（用 1 湯匙凍開水拌勻）

水 1/3 杯

做法

1. 菜洗淨、去外層老莢；切成適口大小。
2. 蒜頭拍扁、去衣，剁成蒜茸。
3. 燒熱油鑊，爆香薑片及蒜茸，下菜猛火快炒 2 分鐘，加水、收中慢火、加蓋燜煮約 15-20 分鐘至菜脍熟，下鹽調味，下粟粉芡汁兜勻，下火腿拌勻，即可上碟。

金菇杞子炒菠菜

富含纖維素
十二朝前
剖腹產者亦宜
全素食

功效：纖維素豐富，有助保持大便通暢。全素。

服法：天天吃也無妨。

宜忌：生薑只是用於起鑊，不會食用，因此對剖腹產者也不會有
影響。

材料

金菇約 200 克
蒜茸 1 湯匙
杞子 12 粒
菠菜約 500 克
生薑 2 片
生油 1/2 茶匙
水 1/3 杯

調味料

生抽、生粉各 1/2 茶匙
麻油 1 茶匙
鹽 1/3 茶匙

做法

1. 菠菜洗淨，去根、切成 3 吋長段。
2. 金菇去根部，洗淨。
3. 燒熱油鑊，下生薑起鑊，下蒜茸爆香，下菠菜及金菇以猛火
   快炒 2 分鐘，收慢火、加水、蓋上鑊蓋煮約 15 分鐘至菜及菇
   煮熟，下已拌勻的調味料及杞子兜勻，炒 3 分鐘，即可上碟。

杞子

滋補肝腎，對產後體虛乏力的產婦有
調補之效。選外表飽滿及顏色暗紅的
為佳。

梅子排骨

增進食欲
十二朝前
剖腹產者亦宜

功效：開胃、醒胃。是很好的餸，對胃納悶者，有增進食欲的功效。

服法：天天吃也無妨。

貼士：可在本港的傳統醬園選購其自家醃製的酸梅。酸梅通常分甜酸及鹹酸兩款，若採用了鹹酸梅，可適當將糖的用量增加。

材料	醃料
有機蔗糖 50 克	生抽 1 茶匙
醃酸梅 3 粒（大）	老抽 1.5 茶匙
一字排骨約 300 克	白胡椒粉、生粉各 1/3 茶匙
生薑 3 片	米酒 2 茶匙
蒜子肉 2 瓣	

做法

1. 酸梅去核，壓成醬。
2. 排骨洗淨，切件，汆水 2 分鐘，撈起，瀝乾水、待涼；加入已拌勻的醃料及酸梅醬，放入雪櫃中醃 30 分鐘。
3. 將所有材料放入鍋內以中大火煮滾，改文火煮 30-40 分鐘即成。

乾蔥炒雞柳

溫和補虛
十二朝前
剖腹產者只用薑起鑊，下雞前取走即可

功效：鮮香惹味，是溫和的產後補虛菜式。

服法：每週 2-3 次。

宜忌：外感者不宜。

貼士：下少許老抽可為雞柳上色，增進食慾。

材料

乾蔥 4 粒
新鮮雞柳約 300 克
薑 3 片
蒜頭 3 粒
葡萄核油 1.5 茶匙
紹酒 2 茶匙

醃料

老抽、生抽、米酒、砂糖各 1/2 茶匙
鹽、白胡椒粉、麻油各 1/3 茶匙
生粉 1 茶匙（用少許凍開水拌勻）

做法

1. 雞柳洗淨後去皮、去筋，挑去雞肉下的白色筋膜，切成適口大小。
2. 將所有醃料拌勻，醃雞肉半小時。
3. 乾蔥、蒜頭拍扁，去衣，切碎。燒熱油鑊後下乾蔥、蒜頭爆炒 2 分鐘，下薑片炒 1 分鐘，灒酒後加鑊蓋緊蓋 20 秒，轉中大火下雞柳兜勻快炒 3 分鐘，改中火炒至雞柳熟即可熄火上碟。

鮑魚炆冬菇

滋陰補虛
十二朝前
剖腹產者亦宜

功效：滋陰補虛，香甜惹味。

服法：每週 1-2 次。

宜忌：外感者不宜。

材料

日本花菇 3 隻（大）

罐頭鮑魚 1 隻（大）

蔥 1 棵

蒜頭 2 粒

木魚汁 2 湯匙

鮑魚水半碗

芥花籽油 1/2 茶匙

調味料

米酒 1 湯匙

老抽、生抽各 1.5 茶匙

冰糖 20 克

生粉 1 茶匙

（用 1.5 湯匙凍開水拌勻）

做法

1. 全部材料洗淨。花菇浸軟後、清去雜質，去掉硬蒂；再用凍開水浸 1 小時，切成適口大小，浸菇水保留。

2. 蔥去根部，蔥綠切成蔥花，蔥白切段。蒜頭拍扁去衣。

3. 鮑魚切片備用。

4. 煲內下油爆香蒜頭後，下蔥白炒香，加入花菇兜炒 5 分鐘，下浸菇水、木魚汁及鮑魚水煮至水滾，改文火將花菇煮至腍熟，下鮑魚及調味料入煲內拌勻，改中火同煮至汁收乾，下蔥花拌勻即可上碟。

西蘭花芝士煎龍脷柳

有益味美
十二朝前
剖腹產者去薑改用蒜頭即可

功效：蛋白質豐富，有益美味。適合所有體質者。

服法：天天吃也無妨。

宜忌：對食材過敏者慎服。

材料

西蘭花 300 克
芝士碎 1.5 湯匙
牛奶 2 湯匙
龍脷柳一塊約 300 克
蒜頭 3 瓣
生油 1 茶匙
檸檬汁 1.5 茶匙
鹽 1/4 茶匙

醃料

鹽 1/3 茶匙
生抽、麻油各 1 茶匙
砂糖、胡椒粉各 1/3 茶匙
白酒 2 茶匙
生粉 1 茶匙
檸檬汁 1.5 茶匙

做法

1. 龍脷柳放在雪櫃下格徹底解凍，沖水；檸檬汁搽勻魚柳，醃 5 分鐘，辟去「雪味」。下已拌勻的醃料醃魚 30 分鐘。

2. 注牛奶入一小鍋內，加入芝士碎及鹽，以文火煮至芝士全溶，邊煮邊攪，以防黐底。

3. 蒜頭拍扁、去衣，剁成蒜茸。

4. 洗淨西蘭花，切成適口大小；用水灼熟，瀝乾水備用。

5. 燒熱油鑊，下蒜茸爆香，下魚柳煎至兩面熟透，下已灼熟的西蘭花炒勻，上碟，淋上芝士汁即成。

鹹蛋黃肉碎炒南瓜

甘香可口
十二朝前
剖腹產者亦宜

功效：美味香口，是葉菜以外補充纖維素的好選擇。

服法：每週 1-2 次。

貼士：可按個人口味選用雞茸或牛肉碎代替豬肉碎。

材料

鹹蛋黃 1 隻
免治瘦肉 200 克
南瓜 300 克
蒜頭 3 瓣
生油 1/2 茶匙
水 1/3 杯
葱花 2 湯匙

醃料

生抽、生粉各 1/2 茶匙
米酒、麻油各 1 茶匙
胡椒粉 1/3 茶匙
水 2 茶匙

做法

1. 南瓜去皮及籽，切成適口大小。
2. 蒜頭拍扁、去衣。
3. 將免治瘦肉拌入已拌勻的醃料中醃 30 分鐘。
4. 鹹蛋黃壓爛，加少許水拌勻備用。
5. 燒熱油鑊，下蒜頭爆香，下肉碎炒至六成熟，離鑊備用。
6. 原鑊下南瓜快速兜炒 3 分鐘，加水、加蓋以文火煮 15-20 分鐘，下已炒過的肉碎炒至全熟，下鹹蛋黃炒勻至全熟，下葱花兜勻即可上碟。

阿膠蛋蜜

止血、治虛損
十二朝後
剖腹產者去薑改用陳皮即可

即時睇片

功效：補血、止血，治虛損。惡露量多，過久不收者宜多服。

服法：每週 1-2 次。

宜忌：外感者不宜。瘀阻沖任者宜先諮詢中醫師意見。

貼士：在藥店買到的阿膠多是一磚磚或一塊塊的，在家烹調時會較難處理，可以請相熟店家，將磚或塊狀阿膠打成粉狀，方便烹調。市面上亦有阿膠粉售賣，但要小心質量，宜幫襯相熟或有商譽的店舖。

材料

優質阿膠 1 両
雞蛋 1 隻
紅糖 30 克
水 450-500 毫升
老薑 3 片

做法

1. 阿膠打研成粉狀。
2. 注清水入鍋中，下薑以大火煮滾後，改文火、下阿膠粉及紅糖，邊煮邊攪拌，直至阿膠粉完全溶化，下雞蛋，拌勻、煮至蛋熟即成。

阿膠

阿膠是用驢皮熬製而成膠狀硬塊中藥，以烏黑、光亮透明、無腥臭氣味者最佳。阿膠味甘性平，有補血止血、滋陰潤肺和安胎作用，並有效調理因出血引起的貧血癥狀。

老薑汁撞奶

祛寒暖胃
十二朝後
剖腹產者不宜

功效：祛風、祛寒。美味暖胃。

服法：每週 1-2 次。

宜忌：內熱盛者慎服。

材料

老薑汁 1/3 碗

優質鮮奶 300-400 毫升

有機原蔗糖 1/2-1 湯匙

做法

1. 將鮮奶注入鍋內，以中火煮滾，下糖煮至糖溶。
2. 在湯碗中放入老薑汁，牛奶離火後即快速沖入薑汁中。
3. 靜置至牛奶凝固後，即可享用。

素食養生薑醋

功效：活血，祛風，有助產後身體恢復健康。全素用料，素食者亦可食用。

服法：份量夠產婦個人常規進食約 15-20 次，天天喝少許醋，對身體最有益處。

宜忌：外感時不服，內熱盛者只宜少量食用。體寒者最好用醋來拌飯食用。

貼士：產前一個月開始預備。每一星期煲滾一次，以免細菌滋生。在瓦煲底墊上竹墊可避免材料黐底。竹墊在用前要先洗淨並在熱水中略煮。嚴格茹素不能吃蛋者，要加黃豆 2 斤同煮，以增加蛋白質的攝取量。

材料

麵筋 2 斤
添丁甜醋 5 公升
去皮老薑 5 斤
雞蛋 15-20 隻
黑糯米醋 500 毫升
片糖 300 克
鹽 1 湯匙

做法

1. 麵筋洗淨、汆水、瀝乾水分，切件，擠乾水分；用白鑊煎 10 分鐘，盛起備用。
2. 雞蛋焓熟去殼。
3. 老薑、拍鬆、切件，加鹽醃 1 小時，沖水，瀝乾水分。瓦煲底放一塊已洗淨的竹墊，將所有材料放入瓦煲內；加醋至浸過材料面，以大火煮滾後，改文火煮麵筋至腍，熄火，趁熱、下雞蛋浸泡。（至少 5 小時）

糯稻根防風大棗茶

益氣止汗
十二朝前
剖腹產者亦宜

功效：益氣止汗，斂汗改善毛孔疏鬆的情況。適合產後數天仍
　　　自汗、盜汗、大汗淋漓不止的媽媽，亦適合怕風、畏寒
　　　者飲用。

服法：每天或隔天一次，直至見效為止。如服用數次後，效果
　　　不理想者，應向中醫師求診。

宜忌：對自己體質有懷疑者，服前請諮詢醫師意見。

材料

炙北芪 3 錢

糯稻根、浮小麥、防風各 2 錢

大棗 4 粒

陳皮一小塊

片糖適量（隨意選用）

水 1800 毫升

做法

1. 全部藥材放在密籮笥箕內，用水沖洗片刻；加水浸 15 分
　 鐘。

2. 全部材料（糖除外）以大火煮滾，改中慢火煮 40 分鐘，
　 隔渣取茶，加糖調味即成。

糯稻根 ● 防風
糯稻根益胃生津，養陰退熱，止盜汗。
產婦在坐月期間出現盜汗（即入睡時
自然出汗），若皮膚把汗濕回吸進身
體，容易引致寒邪入身。防風有發表
散風、勝濕止痛等功效，配北芪同用，
可治表虛外感自汗。

天麻白芷石菖蒲桂圓茶

祛頭風、止頭痛
十二朝前
剖腹產者亦宜

功效：祛頭風，止頭痛，改善頭脹不適。適合產後易頭痛，易頭暈
及頭皮有發麻感的媽媽作日常預防及保健茶水。

服法：每天或隔天一次，直至見效為止。如服用數次後，效果不理
想者，應向中醫師求診。

宜忌：對自己體質有懷疑者，服前請諮詢醫師意見。

材料

薑汁製天麻、白芷、石菖蒲各 2 錢

桂圓肉 3 錢

川芎、炙甘草各 1.5 錢

陳皮一小塊

片糖適量（隨意選用）

水 1800 毫升

做法

1. 全部藥材放在密窿笣箕內，用水沖洗片刻；加水浸 15 分鐘。
2. 全部材料（糖除外）以大火煮滾，改中慢火煮 40 分鐘，隔
   渣取茶，加糖調味即成。

石菖蒲

石菖蒲味辛、性微溫。功效包括：化痰
開竅、化濕行氣、祛風利痹、消腫止痛
等，可有助舒緩產後不適。

利水去腫修身茶

利尿去腫
十二朝前
剖腹產者亦宜

功效：去水腫、利尿，助產後改善身體胖腫，減少面目浮腫、四肢腫脹，以及幫助去水排尿，預防膀胱發炎，特別是因剖腹生產而需插尿管的媽媽，在生產後的數天內是需要多排尿的，此茶有助減少患尿道炎的機會。

服法：每天或隔天一次，直至見效為止。如服用數次後，效果不理想者，應向中醫師求診。

宜忌：對自己體質有懷疑者，服前請諮詢醫師意見。

材料

北芪、澤瀉、赤小豆各 3 錢
雲苓皮、茺蔚子、太子參各 2 錢
炙甘草、陳皮、紫蘇木各 1 錢
水 1800 毫升

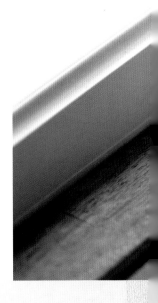

做法

1. 全部藥材放在密窿笪箕內，用水沖洗片刻；加水浸 15 分鐘。
2. 全部材料以大火煮滾，改中慢火煮 40 分鐘，隔渣取茶即成。

茺蔚子 ● 紫蘇木

茺蔚子有活血調經、清肝明目的功效，可舒緩婦女產後瘀滯腹痛。紫蘇木的功效包括：行血、破瘀、消腫、止痛。針對產後惡露不盡有顯著功效。血虛無瘀者不宜服用。孕婦忌服。

淡豆豉韮菜麥芽茶

特效回奶
如有需要回奶隨時可喝
剖腹產者去薑改用陳皮即可

功效：回奶，消乳脹。

服法：分多次飲用，直至效果滿意為止。

宜忌：對自己體質有懷疑者，服前請諮詢醫師意見。

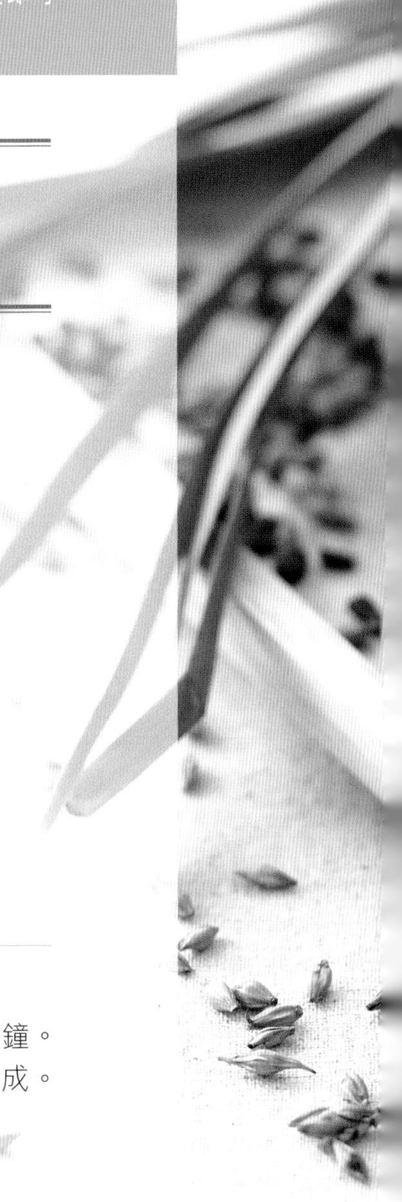

材料

淡豆豉 1 両
韮菜半斤
炒麥芽 1 両
水 2000 毫升
生薑 1 片

做法

1. 韮菜洗淨，切成寸許段長。
2. 全部藥材放在密窿筲箕內，用水沖洗片刻；加水浸 15 分鐘。
3. 全部材料以大火煮滾，改中慢火煮 40 分鐘，隔渣取茶即成。

淡豆豉

產婦生產後如感胸部脹痛,又不餵哺母乳者,可飲用此茶作回奶之用,以減低胸部脹滿不適。

115

清宮祛瘀茶

祛瘀清宮
十二朝後適用

功效：祛瘀清宮，有助子宮收縮，排出殘餘惡露，改善因宮縮不良而引起的輕微腹痛。

服法：1 日內分次服完。隔天再服一劑。

宜忌：本身有易流血不止傾向者，服前宜諮詢中醫師。

材料

艾葉、甘草、蘇木各 2 錢
山楂 3 錢
陳皮一小塊
水 1000 毫升
生薑 2 片

做法

1. 全部藥材放在密窿笒箕內，用水沖洗片刻；加水浸 15 分鐘。
2. 全部材料以大火煮滾，改中慢火煮 40 分鐘，隔渣取茶即成。

第五章

坊間一些以訛傳訛的坐月謬誤，令一般人對產後的調理產生不少誤解。本章以正確的醫學理論去解釋一些坐月習俗，用中醫基礎理論去說明正確的坐月禁忌事項及其重要性，破除謬誤。

坐月迷思的闡釋

親友在「十二朝」前絕對不能來訪？

古時的用意是產後十二日內，不要有訪客。中醫書籍有云：「產後要避風、避客」正是此意。當過了十二朝後，訪客才可到訪。其實十二朝這概念亦很科學，因為產婦在產後應多休息、減少活動，有利傷口快速癒合，太多訪客，除了影響產婦好好休息外，也會令其情緒波動。

此外，新生兒也不宜跟太多訪客接觸，以免令之受驚或被傳染疾病。所以一般在十二朝後才派薑醋給親友，才歡迎至親好友到訪。

在醫學上，「十二朝」的重要性在於中醫認為十二朝後產婦可以進食多一些滋補的食材，以促進復原。

坐月時是否絕對不能外出？

產婦在坐月時真的不應隨意外出，用意是要產婦多加休息及與新生兒多些時間在一起。新生兒每兩小時就要吃奶，媽媽自然不可以離開得太久，而且，新生兒的抵抗力仍是弱的，也不宜多外出，以免感染疾病或受驚，因此，傳統上就讓產婦少外出。

另外，外出也要看天氣情況，如是下雨、下雪、大霧、大風、天冷、太炎熱的日子，除非是十萬火急的事情，否則千萬別外出。即使是常規檢查，也應盡量延期，另約時間。中醫也不鼓勵產婦坐月時在晚上外出，免受陰寒。如果是風和日麗、好天好時的白天，產婦在坐月期間如有需要，是可以外出一段短時間的。中醫學認為日為陽、夜為陰，所以在晚上陽氣不及白天，產婦會易被陰氣所侵，要外出最好是在太陽仍未落山前，天地之間陽氣充足之時，產婦沒那麼容易感受風寒，因而得病。

產後婦女元氣會大傷，多休息有助身體復原，最重要是在產後的首兩周，產婦易感到體力不足，多留在家中休養，對健康是必要的。筆者很反對某些人說：「中醫主張產婦在坐月時絕對不要『去街』」，「去街」之定義是甚麼？平素比較活躍、愛逛街的產婦，請在坐月期間減少非必要的外出，等「滿月」後才「去街」吧！筆者認為如果產婦的身體需要，要找中醫或西醫調理及治療，是應該及時在做好保暖、防風的措施下外出進行的，不

要以為「坐月不應外出」就等到「滿月」後才進行，這會令你錯過了治療及調理的最佳時機。筆者在臨床上常碰到一些產後病人，在「滿月」後才到診所找我治療母乳不足的問題，她們已經錯過了治療及催乳的黃金時間了，即使用上相同的治療方法，很多時療效也未如產後一周就求醫的產婦般理想。筆者常跟這些病人打趣，為何西醫囑咐妳們去覆診，妳們不會說要「坐完月」才可外出而「未去住」？是否覺得看中醫就會被醫師責怪妳們坐月時外出？還是自己誤會了坐月期間連看病、調理身體也不應外出？

筆者給所有產婦的建議是：應酬、逛街等交際活動，別去；但當對自己身體、對健康有好處時，則應在做妥保護措施的情況下外出，別耽誤調養身體的時機。

產婦外出前一定要做好保暖、防風的措施，即使在夏天，也要帶備外套、圍巾等，以抵禦「冷氣」，因為某些室內空調的溫度對新產婦來說有可能太低，而令其受涼。不論是在交通工具上還是診所，也不要對着空調出風口坐，以免寒邪入體。出了汗時要馬上抹汗，內衣即使是稍濕也應立刻更換，切勿讓皮膚把汗濕回吸進身體。到人多擠逼之處，最好帶上口罩，保護自己免受疾病感染。

產婦在坐月期間外出時之準備

- 穿好保暖衣物。
- 帶帽，因頭為「諸陽之首，風先授受」，切勿受寒、風、濕等邪所傷，否則難治。
- 用頸巾保護咽喉，以免肺部着涼，否則易咳嗽。
- 即使是夏天，也別穿短袖衫、短褲或裙子，以保護身體，免被「六邪」從皮膚毛孔入侵身體。中醫説的六邪或六淫是指自然界的各種致病物質，包括：風、寒、暑、濕、燥、熱。新產婦在坐月期間因身體未完全復原的關係，是會較常人易被六淫所傷的。
- 春天多風時，要盡量減少皮膚受涼的面積，大風之處勿去，等車時風大，也應留在室內等，到車來時才出去並馬上上車，勿開車窗吹風。
- 夏天太陽猛時要打傘，以免受暑熱所傷。要穿薄的長袖衣服及長褲，一定要穿襪。
- 夏天即使炎熱，也別對着風扇或空調吹風。
- 秋天時氣候較乾燥，香港又不冷，外出時只要做足防風措施的話，問題不大。
- 冬天天氣冷，最好少外出，如需帶 BB 去做檢查或打針，自己又要看醫生 / 醫師的話，最好安排在同一天進行，減少外出的次數及時間。一定要帶帽及手套。

產婦在坐月期間切勿進行的活動

- 不宜到醫院探病，以免被傳染疾病。
- 不要去人多擠迫及空氣不流通的地方，以免被傳染疾病。
- 不要去美容院做皮膚護理，不要做水療、按摩，不要去髮型屋洗頭，因那些場所用的均是生水。
- 坐月期間不宜浸浴。
- 不宜過於頻密地洗頭。
- 不要去海邊或沙灘吹風，不要游泳。
- 不宜穿很少衣服來曬太陽。

產後不能洗頭及洗澡？

　　筆者在教授陪月員中醫知識時，常被問到產婦在坐月期間，是否一定不可以沖涼及洗頭。也被自己的婦產科醫生朋友説笑地問及，中醫是否叫產婦不要沖涼和洗頭？現在正好趁這個機會澄清這些謬誤：中醫從來都認為坐月期間一定要注意個人衛生的，沒有不准產婦洗澡及洗頭的規條。

以前傳統上會說產婦不宜洗澡及洗頭，筆者對此的理解是因為從前的家居設施較為落後，衛生條件亦沒有現在的好，要進行沐浴洗髮等活動，要擔水、煲熱水及用大盆來裝水，更沒有電暖爐、風筒等設備，產婦洗頭、沐浴容易着涼生病，或引起頭痛、關節痠痛等問題，因此，很多產婦就不在坐月時洗澡、洗頭。這是歷史因素。

即使中醫學也是與時並進的，今時今日，特別在香港，家居設備良好，衛生條件大大改善，只要依照以下辦法、規則去沐浴及洗頭，對身體健康之危害是不大的。

坐月時沐浴及洗頭注意事項

- 勿用生水、冷水洗手、洗澡及洗頭。接觸身體的最好是用熱薑皮水，水溫應以個人可接受的最高限度為限，但也要小心燙傷。薑皮也不要下得太多，以免水「太辣」引起皮膚敏感及灼傷，對薑過敏者，可以用艾葉代之。
- 自然分娩者，產後當天只宜以熱薑水抹身，最好不要洗澡。
- 剖腹產者手術後 3 至 4 天內也只宜以熱薑水抹身，最好不要洗澡。
- 如非必要勿天天洗澡，特別是在寒冷的冬季。
- 洗澡時間要短，浴後馬上擦乾身體，穿足夠衣服保暖。
- 注意浴後的傷口護理。
- 應勤換衛生棉，以免細菌感染。
- 產後 5 至 7 天內如非必要，切勿洗頭。
- 洗頭要快速，洗頭後迅速將頭髮用熱風吹乾。
- 避免在當風的地方進行洗澡及洗頭，也別在清晨和晚上等氣溫較低時洗頭沐浴，就不容易惹風寒了。
- 氣溫低時最好使用暖爐、暖氣來令室內氣溫保持溫暖，但切勿用暖風機，即使是暖風，產婦也不宜被吹到。
- 天氣潮濕時要用抽濕機，將室內的濕度降低，產婦被濕邪所傷的機會就會降低。
- 在生產過程中有異常狀況，或產後復原欠佳者，請先諮詢醫生 / 醫師意見才進行。

坐月時不能開空調及風扇？

老實說，有許多傳統習慣是不科學的，筆者身為中醫、也曾是助產士，在行醫及從前在醫院工作時，也看到、聽過很多奇怪的坐月習俗。例如：坐月期間，為怕着涼及被風侵害，產婦在坐月時不論天氣一定要穿厚衣服、戴厚帽、關門、不開窗、不吹風扇、不用空調，為免腸胃受寒，不吃生果、蔬菜等等。

筆者的角度是如何穿衣應以天氣為依據，冷天當然是要好好保暖不要受涼，相信即使不是坐月的產婦也應如是吧！不過在夏天坐月時，筆者還是要提醒產婦，最好別將四肢暴露出來，更不應穿背心、短褲、露臍裝等，應穿透氣乾爽的長袖衫褲。產後一般頭數天甚至十天，產婦均會較多汗液排出，如稍不慎，未有及時抹乾身體及更衣，的確很容易感染風寒，也很易被汗濕所傷而致病，產後體弱又加上患病，是較一般人難治理及更覺辛苦的。因此，內衣褲絕對要保持乾爽，不能掉以輕心，一旦有汗濕，馬上要更衣，若在夏天，更衣、沐浴及洗頭時要把空調及風扇關掉，等穿好衣服、頭髮乾了才再開空調或風扇。

在香港這種炎熱的夏天時坐月，如果不開窗透氣、不開空調、不用風扇，反而是會被「熱到病」的！試想像一下，天天30多度的氣溫，人又要留在室內，像不像焗桑拿？天天這樣發汗，想不虛、不病也難了。

產婦在坐月期間使用空調及風扇的注意事項

- 不要將氣溫調得過低，應以攝氏23-25度為宜。風速也不應太大。不要將出風口對着產婦。
- 風扇的確是可用來生風，但因為要對着風扇吹才會覺得涼快，產婦很容易在吹風扇時感受風邪而不自知。如果真的要用風扇，切勿對着坐月的產婦直接吹，應將風扇向着相反方向吹，令空氣流動而帶來涼爽的感覺。
- 筆者認為溫度不過低、風速不強的空調比風扇好。

坐月時不可吃生果、蔬菜？

這絕對是謬誤。不論從中醫或營養學的角度來看，生果、蔬菜是一定要吃的，而且可以天天吃。筆者認為只要懂得挑選適合自己體質、知道要避開那些蔬果，坐月期間日日吃蔬果也沒問題。

中醫不主張產婦在坐月期間吃生冷、寒涼的蔬果，例如：芥菜、白菜、葛菜、苦瓜、西芹、綠豆、西瓜、山竹、水梨等。一些溫和不寒涼的蔬果，是應該吃的，不吃蔬果，營養會失衡，也易引起便秘。當補得太多，體內易聚積熱氣，更應以溫和清潤的蔬果來調節，以免火熱聚積而生病。

自然分娩與剖腹分娩在坐月的調理上有何不同？

即時睇片

　　無論是自然或剖腹分娩，基本上都要在產後 12 天才可開始進食滋補效能強的食物，例如：八珍湯、十全大補湯、阿膠、鹿茸等。在正常的情況下，產後 12 天，惡露基本上已很少，傷口理應幾乎癒合了，此時進食大補氣血的食物不會造成出血及影響傷口。

　　自然分娩或剖腹分娩在坐月時的調理原則基本沒有不同，均強調產婦必須要充分休息，再以符合個人體質的食療補身，同時亦要有適當的輕柔運動，不可吃過於寒涼、辛辣刺激、太過燥熱、油膩和難以消化的食物。

　　剖腹產者腹部會有傷口，為避免傷口感染、出血等，在坐月期間，應少吃鵝、鴨、鱔、甲殼類海產等容易影響傷口之食物；另外生魚、山斑魚和薑，多吃是會引起肉芽的，這些食物要待傷口完全癒合後才可進食。蒸魚、煲湯、炒菜時可用少許薑起鑊、辟腥味、去風寒，只要產婦不是大口直接吃薑，是沒問題的。產婦可趁坐月期間多進食補益氣血的食物，幫助改善身體，有利日後的健康。

　　剖腹產者多因為傷口痛而不敢用力排便，加上大部分產後補身食物都較溫燥，容易引起便秘或令痔瘡惡化，因此保持大便易排、暢通是很重要的。必須多吃富含纖維素的食物，多喝水，適當攝取少量健康油分，也有助令大便柔軟易排。不宜多吃過於滯膩、燥熱的補身食療。補身食療應以補、平補、稍清潤的次序去安排。例如：喝了一天雞酒，次日可喝木瓜魚湯，再來可飲八珍雞湯，接着飲清補涼湯。

坐月時是否不能吃紅棗、木耳等具活血功效的食材？

　　看妳怎樣吃、吃多少，以及自己的體質適合否。如果產婦本身是有凝血問題的，產後是不宜吃任何行血、活血及對凝血機能有影響的食材。如果產婦身體一切正常，生產的過程中沒有任何問題出現，產後又一切無異，吃些少紅棗、木耳、雲耳、雞蛋煮酒等是不用害怕的；如紅棗雲耳蒸雞、木耳煮雞酒等，只要食材份量不過多，一般正常情況下是不會引起大出血的。

　　某些產婦的體質是較易有血瘀的，吃了這些溫和行血的食物，是會有一時的惡露增多或有血塊排出，這些現象只要不是大量及持續出現，便不用擔心，這是食材起到療效的表現，對身體反而有好處。

　　總之妳對自己身體有任何懷疑，不清楚可以吃甚麼食材時，最好諮詢中醫師。中西醫的理論基礎是截然不同的，有些對中醫藥不了解的西醫，或會以他 / 她們自己的角度去看中醫的食材，所以便會出現中、西醫不同的解釋，筆者認為若果本身不支持中醫學說、不相信中醫藥療效的人，請勿以中醫學說的主張來進行坐月食療。理論上西醫是沒有坐月及食療藥膳等主張的，應完全遵循西醫的囑咐去做。

坐月期間可喝酒嗎？

答案視乎妳喝甚麼酒、怎樣喝、喝多少、何時喝？絕不能喝冷酒、凍酒；不宜喝太多、太密；最好是喝煮過的酒，如豬腰雞蛋煮糯米酒、黃酒煮雞等；不能在餵奶前喝酒。中醫認為適量的酒有行血行氣、祛寒去濕的功效，但不是叫妳直接將酒灌進肚子，是很講究烹調的。一般主張用適量（大概 200-400 毫升）的米酒或糯米酒，配搭其他食材如雞、豬腰、雞蛋、糯米丸子等，一起經過烹調程序，以求去除酒精的酒性才給產婦吃的。經過烹煮的少量酒，是不會令你喝醉、也不會令寶寶吃人奶後也醉倒，更不會引起大量出血的。

是不是所有產婦都要在坐月期間進補？

並不是所有產婦都要進補的。精力充沛、氣血旺盛的產婦，只宜平補、清補，不要過補、大補。產婦在患病時，須暫時停止進補，應先延醫診治，治理好疾病才跟隨醫生／醫師的指示進補。

進補的首要原則是要「對證用補」，即是要根據產婦個人體質之特性去處理，決定吃甚麼食材。宜慢慢進補，忌操之過急，補得太急太多，療效不佳之餘還會引起不適。要根據個人的消化能力、吸收能力去進補。

進補的意思也不只是一味吃昂貴的補品，很多普通的食材只要運用得宜，也可達到補身的功效，不要看輕淮山、蓮子、玉竹、沙參，雪耳等便宜食材，它們都能有助產婦復原，而且性質較為平和，幾乎人人都受得。

一般產婦體質多虛，多瘀。因為生產時氣血耗損，津液虧虛，產後如再加上飲食不節，很易損傷脾胃，引起脾病。而瘀是由於產程之中瘀血內阻，敗血妄行引致。另外，虛又可以致寒，所以產婦易得虛寒證候，食療宜辛溫為主就是這個原因，辛溫之品有助祛寒補虛，但也不宜多服太過溫燥之品。

產前多火，不必大補，產後虛寒多，是進補的最佳時機，補要補身體不足，故此，一定要掌握好體質的密碼才進補。

看到奶粉廣告説奶粉有多種添加成分，能加強嬰兒抵抗力、促進腦部發展等等，人奶的營養成分怎能及得上奶粉？

大家必須先明白一點：奶粉生產商研製奶粉，必然是以模仿人奶成分為目標。餵哺母乳是哺乳類動物與生俱來的天賦本能，產婦只要注意飲食、採用正確的方法，必定能透過母乳提供充足的養分給她們的嬰兒。

產後餵母乳的媽媽，為了有足夠的奶水，除了保持均衡飲食外，還需要比平時攝取更多的高蛋白質。產婦在產後要適當地多喝湯、多喝水、牛奶等「補品」；也應比平時多吃一至兩餐，以促進乳汁分泌，令乳汁的質量更好。

授乳期媽媽不能不謹慎挑選自己的飲食，因為妳吃喝甚麼，都是會直接影響 BB 的。有些過度進補的產婦，在寶寶喝了一段時間母乳後，會出現不同的熱氣表現如：便秘、大便乾硬難排、皮膚出紅疹、嘴唇周圍生皮疹、皮膚長瘡、夜睡不寧甚至啼哭等。

餵哺母乳的媽媽要注意的事項

- 不能吃會抑制乳汁分泌的食物，例如：麥芽、韭菜、豬膶、淡豆豉等。
- 產後飲食宜清淡，不要吃辛辣、刺激性的食品，如：過於辛辣的調味料、酒、咖啡及火鍋、燒烤、油炸的食物。
- 不要吸煙。
- 小心吃藥。不論中西藥均不要自己找來吃，服藥前請先諮詢醫者或專業人士。看病時也要主動告訴醫生自己正在餵乳，以便醫生開出適合的藥物，並選擇半衰期較短的藥物，以防止過多的藥量通過乳汁被寶寶吸收。
- 小心觀察嬰兒有否出現過敏情況。新生兒有可能對某些物質過敏，媽媽要多留意寶寶皮膚的狀況，觀察有否出現紅疹，並評估自己的飲食，以盡早發現問題所在，找出食物的關連性。餵母乳者，應避免吃任何有可能會引起寶寶過敏的食物。

第六章

婦女產後身體經常會出現一些生理現象的轉變，不但對產婦的健康構成影響，嚴重的更會為她們帶來困擾，影響情緒。本章就一些常見的現象作出解答及提供一些對策，希望能助產婦輕鬆面對。

產後為何會易脫髮？

據統計，約有 35％～ 40％的產婦有產後脫髮的現象。產婦常見的一種情況是休止期脫髮。休止期脫髮與產婦的生理、心理、精神等因素，以及其飲食、生活方式有着密切的關係。妊娠期分泌的雌激素較平時增加，令頭髮的生長期延長，脫髮的速度變慢，大量應代謝而未掉的頭髮留在頭皮上，等到產後雌激素水平急速下降後，這些應掉而未掉的頭髮就會驟然消失，令人以為產後出現脫髮的問題。

筆者很多病人都有這現象，一般只需放鬆心情、多休息、轉用溫和、不刺激的洗髮用品後，情況都有改善。產後體虛，需要進補，但很多人認為若產後脫髮則更應猛補、大補、強力地補才是，筆者不同意此觀點。不當的進補，會造成熱盛、濕熱、燥熱等，對健康反而有害。建議經常以木梳梳頭，或用手指在頭皮上作有節律的叩敲、按摩，可以刺激頭皮，促進頭皮的血液循環，幫助頭髮的新陳代謝，有利生髮。

中醫認為「髮為血之餘」，「腎主骨，其榮在髮」，生產期間耗氣失血，血虛津虧會造成脫髮，或出現頭髮枯黃、開叉、無光澤、幼弱等現象。產後若精神壓力過大，情緒憂鬱，令肝氣不舒，也會有掉髮增多的現象。筆者認為產後飲食必須均衡，多補充蛋白質，且睡眠要充足，也要保持心情愉快，同時可以用中藥的藥膳來調理身體、以茶飲來治理脫髮。正常人每天約會掉落 50~100 條頭髮，而產後掉髮問題，若處理得當，通常會在產後半年內改善。

治理產後脫髮，以固髮為原則，內服中藥，以滋陰養肝腎，調補氣血，再輔以促進毛髮生長的藥材，以養血祛風為本。可多食黑豆、黑芝麻、核桃、淮山、菠菜、紅蘿蔔、黑木耳、韭菜、黑棗、桑椹、杞子、花生、栗子等。

以下均是中醫用來治療脫髮的常用中藥材：製何首烏、菟絲子、補骨脂、枸杞子、當歸、山藥、丹參、黑豆、黨參、北芪、陳皮、桂圓肉、炒酸棗仁、桑椹子、炙甘草、大棗、白芍、熟地、阿膠、天麻等。

同時避免進食油膩、辛辣食物，減少頭油分泌。要保持樂觀、心情舒暢，要有充足睡眠，一段時間後定能改善髮質及脫髮情況。另外，有些嚴重脫髮、甚至斑禿者，建議暫停授乳，待身體把精血保留在改善體質上，防止問題變得更嚴重。

適度的頭部按摩可以促進頭部血液循環，幫助頭皮毛囊的新陳代謝，讓頭髮恢復生長。用手指腹由後腦根部向着前額髮際處來回輕敲按摩約 5 分鐘，令頭皮微有發熱感為佳。早晚各做一次即可。按摩手勢、動作要輕柔，而且千萬不要以手指甲抓弄頭皮，以免刮傷頭皮，造成感染，反而掉髮更多。

產後脫髮按體質可分三種類型

- 腎虛型：治療以補腎生髮為主。
- 脾虛血虧型：治療以健脾養血生髮為主。
- 陰血虧虛型：治療以益氣養血生髮為主。

一些非中醫對產後掉髮的看法：

- 分娩後人體內的雌激素明顯下降，導致毛髮自然脫落。
- 產後因為家務煩瑣辛苦，照顧新生兒也要適應，往往令產婦感到身心疲累，影響身體的代謝功能，令頭皮血運不暢，令頭髮變得幼弱。
- 睡眠不足也是新產婦的掉髮原因之一。
- 有些產婦不注重頭部衛生，不洗頭，甚至少梳頭，結果導致頭皮聚積油脂、髒物，甚至引起頭皮發炎或毛囊感染等，使毛髮自然脫落。頭皮敏感也可能引致掉髮。

產後臉上、身體某些部位出現色素沉着又如何處理？

有些孕婦在懷孕後，臉上的顴骨、眼眶及口唇周圍會出現邊緣不規則的棕黑色斑塊或色素沉着，幸運者這些礙眼的色澤會在產後自動消退或減淡，而不幸的那些不雅的色素即使生產後也不會消減。有些婦女因懷孕而出現色素沉着的身體部位，未必是別人可以隨便看得見之處，許多孕婦會在乳頭、乳暈、腋下、會陰、大腿內側等的皮膚出現色素沉着，以黑色及深褐色最為常見。原因是懷孕期間，體內荷爾蒙變化而造成皮膚形成黑斑或局部皮膚顏色變深，可稱之為「孕斑」。

從中醫的觀點來看，皮膚膚色黯淡無光，甚至產生色斑，多與肝、脾、腎有關，其中以肝的影響最大。肝氣必須條達，肝也與情志、情緒等心理有關，懷孕時若壓力過大、精神緊張，或是產後有憂鬱、惡露排出不暢、身體恢復不良、情緒不佳等情況，均可導致肝氣鬱結，氣血不暢，日久表現於皮膚上，就是表面所看到的色斑、斑點及色素沉着了。產後婦女可以通過調理身體來改善以上問題。一個很有用的提示就是一定不要曬太陽，否則，皮膚上的那些色素就好像移民到你的身上一樣，不會離開了。

中醫一般都會按個人體質去辨證論治，為患者選擇治療的外用或內服藥，也會提供食療建議，令妳喝出健康好氣色。最重要是產婦本人一定要保持好心情，不要發「忟憎」，肝火盛時，面上的色斑是不肯走的，而且還會愈長愈多。

適用於產婦的美容祛斑藥材，一般包括：枸杞子、當歸、黨參、北芪、紅棗、杜仲、何首烏、白芍、白芷、覆盆子、桑寄生、素馨花、鬱金、浮小麥等。取其補氣養血、柔肝養肝、滋陰等功效來解決色素問題。

產後為何會容易肌肉疼痛？

在妊娠期間，胎兒的發育令子宮增大，腹部的重量增加，孕婦的腰部會承受着較平時重數倍的重量，腹部因變大而向前凸出，令身體的重心亦隨之向前傾，人自然會刻意用腰力使身體靠向後以保持重心適中，對腰背造成很大的壓力，對關節、韌帶、肌肉均帶來影響，亦有可能因為不正確的姿勢、十月懷胎、生產時用力娩出胎兒之動作等，最終引致很多產婦在產後出現腰背部、臀部及腿部痠痛。

有些產婦由於生產時用錯力，或會弄傷腰骨和盆骨，引起腰、背或臀部疼痛。腿部則是因為分娩時的姿勢、緊張，雙腿長時間抬高置於產床的腳蹬上，令雙腿一直很不自然地擱着，肌肉有可能被拉傷，引起腿痛。

產前預防腰背痛要點

- 孕婦應使用腰封，幫助腰背支撐腹部的重量。
- 懷孕期間應保持正確的坐、立、躺、卧等姿勢。
- 切勿彎腰提重物。

產後筋骨、肌肉護理要點

- 不要長時間抱嬰兒。
- 要學懂收緊腹部，挺起胸部才抱持較重物件；提起重物時要先屈膝利用雙腿發力，千萬不可彎腰，以防拉傷腰部肌肉及弄傷腰部關節。
- 保持正確坐立姿勢，餵奶時要先坐好，用軟墊支撐好手臂及腰背。
- 從床上起來時應先轉身側卧，雙腳放出床邊，以手的力量慢慢支撐起上身才起床，不要突然急劇的起來，也不要長時間維持在同一個姿勢。
- 產後三個月內不能做劇烈運動，以保護腰部，但應做適當的產後運動。
- 穿戴腰封除了可以幫助減輕腰背的壓力外，也可減輕子宮收縮的痛楚。
- 腿部「抽筋」痙攣多在晚上發生，可用熱敷來增加血液循環，減少痙攣；也可按摩局部肌肉，有助放鬆。
- 當產後腰痛或腿痛，又或者關節出現活動不靈時，可試吃杜仲、桑寄生、雞血藤等食療湯水，如不見效，應該就醫。
- 針灸對以上不適有不錯的療效。

產後出現很多不適癥狀，有何對策？

遇有不適，最適當的處理方法當然是延醫診治。但其實大部分癥狀都是產後常見的現象，產婦毋須過分擔憂，以免影響情緒。只要小心處理，這些現象便會慢慢得以舒緩。產後的不適現象大致可以虛證和實證區之：

- 虛證：多指人體正氣（人本身元氣及抵抗力）虛弱者，假使一個人經常容易疲勞、呼吸不順、氣短、乏力、精神不振、自汗（不因天時、衣厚、勞作而白晝時時出汗，動則更甚）、盜汗（寐中出汗，醒來自止）、頭暈、心悸等，多是體虛的人。虛證產婦之正氣不足（抵抗疾病能力不夠），宜用「補法」來增加病人對疾病的抵抗力及令身體復原。

- 實證：多指病邪而言，如：痰多、胸悶煩燥、易腹脹、痛時拒按、便秘等。實證者不宜隨意進補，宜先解決其自身的病邪。

產婦發熱可分為以下三類

- 外感發熱（如出現鼻塞、流鼻水／鼻涕、咳、喉癢／痛、有痰、發熱、怕冷、忽冷忽熱，腹瀉，大便秘結，嘔吐）
- 血瘀發熱（如產後小腹疼痛、拒按、口乾、不想飲水、舌色紫暗）
- 血虛發熱（如產時失血較多、身有微熱、頭昏眼花、自汗、心悸、少寐、腹部隱隱作痛、喜按、手腳麻木）

附錄：產後兩週建議餐單

　　根據筆者行醫及以往任職註冊護士及助產士的經驗，產婦坐月期間的飲食最令人頭痛！尤其是初為人母的首兩週，無論是自然分娩的產婦還是剖腹產者，均會感到身心疲累，甚至出現不同程度的不適癥狀，而且，自己和陪月員定必盡心盡力照顧新生兒，哪有閒情去設計餐單？然而，坐月期間的膳食對於產婦的復原，正正扮演着重要的角色，不容忽視！

　　有見及此，筆者在本部分特別設計了 14 天的產後餐單，供讀者參考。除了包含本書中介紹的補健湯水及養生食譜，也配合一些簡單易做、營養豐富的美食，令坐月的日子飲食不再單調，讓各位新任媽媽保持愉快心情！

　　剖腹產者須注意以下餐單及食譜中的備註，並以其他食物取代有薑的菜式，以免影響傷口癒合。

第 1 天

全日湯水：木瓜牛奶鯇魚濃湯（P.20）

早餐：火腿蛋三文治（麵包可稍烘加溫）＋熱鮮奶一杯

午餐：薑蛋炒飯＋灼菜心＋蒸魚

下午茶：焓蛋一隻＋蘋果一個（中）

晚餐：蒸雞＋炒菠菜＋豬腰煮酒＋蒸大蕉一隻（助排便）＋白飯或糙米飯

宵夜：熱鮮奶麥皮一杯

※ 剖腹產者只喝炒米茶，並遵醫囑進食，若未能吃固體食物，可吃瘦肉粥

第 2 天

全日湯水：椰子紅豆通草鮮八爪魚湯（P.22）

早餐：瘦煙肉芝士粟米湯通粉＋熱鮮奶一杯

午餐：梅子排骨（P.92）＋蒜茸炒紅莧菜＋西椒炒肉絲＋白飯或糙米飯

下午茶：桑寄生蓮子蛋茶＋蘋果一個（中）

晚餐：鮑魚炆冬菇（P.96）＋蒸魚或煎魚＋炒菜心＋白飯或糙米飯＋紅提子 10-15 粒（室溫）

宵夜：燉雙皮奶一碗

※ 剖腹產者只喝炒米茶，並遵醫囑進食，若未能吃固體食物，可吃魚肉粥

第 3 天

全日湯水：蓮子紅豆通草三文魚頭湯（P.40）

早餐：鹹豬骨陳皮粥 + 熱鮮奶一杯

午餐：八爪魚杞子雞蛋飯（P.72）+ 蒜頭炒南瓜 + 蜜糖豆炒雞柳 + 橙一個

下午茶：豆苗豬肉水煮餃子 + 水蜜桃一個

晚餐：蒜茸三文魚扒意粉 （P.74）+ 白灼春菊或皇帝菜 + 荷包蛋一隻

宵夜：鮮奶燉花膠一碗

※ 剖腹產者只喝炒米茶，並遵醫囑進食，應已可進食固體食物，但以易消化為主

第 4 天

全日湯水：珍珠肉石斛海參湯（P.54）

早餐：花生醬牛油多士 + 熱鮮奶一杯 + 炒蛋一份

午餐：松子菜粒肉碎炒飯（P.80）+ 意大利火腿燜津白（P.88）+ 洋蔥煎豬扒

下午茶：熱黑豆漿一杯 + 薑水煮芝麻湯丸 4-6 粒 + 橙一個

晚餐：乾蔥炒雞柳（P.94）+ 陳皮免治牛肉紅棗飯（P.76）+ 番茄炒蛋

宵夜：老薑番薯糖水 + 冰糖煮啤梨一個

第 5 天

全日湯水：黃鱔花生湯（P.42）

早餐：吞拿魚洋蔥包 + 熱鮮奶一杯

午餐：金菇杞子炒菠菜 （P.90）+ 西蘭花芝士煎龍脷柳（P.98）+ 白飯或紅米飯 + 蘋果一個

下午茶：清蛋糕或馬拉糕一件 + 熱黑豆漿一杯

晚餐：薑汁黃鱔薑黃焗飯（P.78）+ 鮮菇荷蘭豆炒雞柳 + 魚滑釀豆卜

宵夜：蒸大蕉一條 + 黑芝麻糊一碗

第 6 天

全日湯水：核桃雪耳無花果響螺湯 （P.44）

早餐：芝士雞蛋奄列 + 熱鮮奶一杯 + 炒磨菇

午餐：栗子冬菇田雞腿紅米飯 （P.82）+ 薑汁酒炒芥蘭 + 乾葱香煎雞塊

下午茶：菜肉包 + 核桃奶一杯 + 橙一個

晚餐：鹹蛋黃肉碎炒南瓜（P.100）+ 荷葉紅棗金針蒸田雞 + 木耳煮豬腰酒 + 白飯或紅米飯

宵夜：酸芝士葱花煙肉碎焗薯一個或薯茸一碗 + 黑提子 10 粒

第 7 天

全日湯水：番茄雞肉大豆芽山楂湯 （P.46）

早餐：叉燒包 + XO 醬炒腸粉一碟 + 熱鮮奶一杯

午餐：紅蘿蔔炒肉絲 + 栗子炆雞 + 清炒豆苗 + 白飯或糙米飯

下午茶：暖芝士蛋糕 + 黑豆漿一杯 + 火龍果一個

晚餐：香蒜排骨冬菜蒸糙米飯（P.84）+ 紅菜頭炒牛肉絲 + 蒸魚

宵夜：果醬牛油多士 + 蒸木瓜一個

第 8 天

全日湯水：冬菇海參鮑魚湯 （P.36）

早餐：奶黃包 + 茄汁焗豆 + 火腿肉絲炒米粉 + 熱鮮奶一杯

午餐：肉醬意粉 + 青豆粟米粒白汁燴魚柳 + 灼莧菜 + 蘋果一個

下午茶：杏仁茶一碗 + 焓蛋一隻 + 豆漿一杯

晚餐：瑤柱蛋白牛肉鬆炒飯 + 紅燒排骨 + 金菇杞子炒菠菜（P.90）+ 車厘子 15 粒

宵夜：椰汁紫米露一碗

第 9 天

全日湯水：金銀百合南瓜素湯（P.48，非素食者可加豬肉）

早餐：皮蛋瘦肉粥 + 熱鮮奶一杯

午餐：健康臘味生炒糯米飯（P.86）+ 鹽水浸菜心 + 雞蛋煮酒 + 橙一個

下午茶：魚湯米線一碗 + 黑豆漿一杯

晚餐：香檸煎雞塊 + 蜜糖豆炒牛肉 + 白飯 + 瑤柱炆節瓜 + 士多啤梨 10 粒

宵夜：燉雙皮奶一碗

第 10 天

全日湯水：蓮子紅豆通草三文魚頭湯（P.40）

早餐：番茄肉絲湯意粉 + 熱鮮奶一杯

午餐：紅莓汁煎羊扒 + 薯茸 + 灼西蘭花 + 橙一個

下午茶：蛋撻一個 + 紅豆沙一碗

晚餐：薑汁黃鱔薑黃焗飯（P.78）+ 薑汁酒臘腸炒芥蘭 + 蒜茸蒸肉餅 + 火龍果一個

宵夜：黑芝麻糊一碗

第 11 天

全日湯水：珍珠肉石斛海參湯（P.54）

早餐：肉絲大豆芽炒麵 + 熱鮮奶一杯

午餐：蒸魚 + 白飯 + 鮮冬菇椰菜炆排骨 + 水蜜桃一個

下午茶：燉蛋一碗 + 黑豆漿一杯

晚餐：黃酒煮雞 + 蒜茸三文魚扒意粉（P.74）+ 意大利火腿燜津白（P.88）

宵夜：花生糊一碗 + 蒸木瓜一個

第 12 天

全日湯水：核桃雪耳無花果響螺湯（P.44）

早餐：玫瑰果醬奶油伴鬆餅 + 熱鮮奶一杯

午餐：金菇杞子炒菠菜（P.90）+ 白飯 + 鮮百合炒雞絲

下午茶：火腿蛋治 + 黑豆漿一杯 + 蘋果一個

晚餐：瑤柱蛋白牛肉鬆炒飯 + 炆花膠海參 + 灼菜心

宵夜：菜肉包一個 + 紅棗蒸啤梨一個

第 13 天

全日湯水：乾榆耳羊藿葉素湯（P.50）

早餐：牛肉粥一碗 + 熱鮮奶一杯

午餐：八爪魚杞子雞蛋飯（P.72）+ 番茄炒蛋 + 白灼
　　　豆苗 + 橙一個

下午茶：阿膠蛋蜜（P.102）

晚餐：素食養生薑醋 （P.106）+ 鹹蛋黃肉碎炒南瓜
　　　（P.100）+ 炆鮑魚雞腳 + 白飯 + 提子 15 粒

宵夜：老薑汁撞奶（P.104）

第 14 天

全日湯水：黑豆川芎天麻鯽魚湯（P.38）

早餐：熱鮮奶一杯 + 楓糖漿熱香餅 + 焗蛋一隻

午餐：松子菜粒肉碎炒飯（P.80）+ 炒菠菜 + 瘦肉炆節
　　　瓜 + 水蜜桃一個

下午茶：南瓜桂圓露 + 豆漿一杯

晚餐：陳皮免治牛肉紅棗飯（P.76）+ 炆津白 + 炒牛肉

宵夜：粟米核桃糊一碗

目錄
Contents
Resep

Papaya, Milk and Grass Carp Thick Soup
Kuah Pepaya, Susu dan Ikan Wang

- within 12 days after giving birth
- can be taken every day
- skip the ginger for women after caesarean
- take less if you do not want breastfeeding

Ingredients:

1.2 kg half-ripe local papaya
800 ml skimmed milk
1 grass carp tail (0.9-1.2 kg)
3 ginger slices
1 small piece dried tangerine peel
5 red dates (pitted)
oil
2000 ml water

Method:

1. Skin the papaya, remove the seeds and cut into pieces.
2. Scale and rinse the grass carp tail. Fry with a little oil until both sides are light brown. Take out and wipe the oil away with kitchen paper. Set aside.
3. Soak the tangerine peel in water until soft and scrape off the pith.
4. Pour water into a tall stock pot. Add the papaya, grass carp tail, ginger, tangerine peel and red dates. Bring to the boil over high heat, turn to low-medium heat and cook for 30 minutes. Pour in the skimmed milk and cook for 20 minutes. Season with salt. Serve.

- dalam waktu 12 hari setelah melahirkan
- melewatkan jahe untuk wanita setelah operasi caesar
- dapat makan setiap hari
- jika tidak mau menyusui, makan sedikit

Bahan:

1.2 kg setengah matang pepaya
800 ml susu skim
1 ekor ikan wang (0.9-1.2 kg)
3 lb. jahe
pojokan kulit jeruk kering
5 bj. tanggal (buang bijinya)
sedikit minyak
2000 ml air

木瓜牛奶鯇魚濃湯

Cara Membuat:

1. Kupas pepaya, buang biji, potong kotak^2.
2. Ikan wang bersihkan sisiknya, cuci bersih. Ikan lalu goreng sampai agak kuning. Pakai kertas dapur menyerap minyaknya.
3. Kuilt jeruk rendam pakai air dingin sampai empuk. Kerok seratnya.
4. Panci besar kasih air. Pepaya, ikan wang, jahe, kulit jeruk, bidara merah, masukkan kedalam panci. Pakai api besar masak sampai mendidih, pakai api sedang masak 30 menit, taruh susu skim, masak 20 menit, lalu kasih garam. Jadilah.

Fresh Octopus Soup with Coconut, Red Beans and Tong Cao
Kuah Kelapa, Kacang Merah, Tong Cao dan Gurita

- within 12 days after giving birth
- take it once every two days
- skip the ginger for women after caesarean
- take less if you do not want breastfeeding

Ingredients:

600 g old coconut pulp
1.8 kg fresh octopus
150 g red beans
75 g Tong Cao
300 g lean pork
3 ginger slices
1 small piece dried tangerine peel
5 red dates (pitted)
2000 ml water

Method:

1. Rinse and gut the octopus. Rub the mucus off with coarse salt. Cut into pieces and blanch in boiling water. Set aside.
2. Rinse the lean pork, blanch in boiling water and cut into pieces. Set aside.
3. Soak the dried tangerine peel in water until soft and scrape off the pith. Pack the herbal ingredients in a paper bag for decoction or in a muslin bag.
4. Pour water into a tall stock pot and add all the ingredients. Bring to the boil over high heat, turn to low-medium heat and cook for 60 minutes. Season with salt. Serve.

- dalam waktu 12 hari setelah melahirkan
- melewatkan jahe untuk wanita setelah operasi caesar
- makan sekali setiap dua hari
- jika tidak mau menyusui, makan sedikit

Bahan:

600 g ampas kelapa
1.8 kg segar gurita
150 g kacang merah
75 g Tong Cao
300 g daging kurus
3 lb. jahe
pojokan kulit jeruk kering
5 bj. tanggal (buang bijinya)
2000 ml air

Cara Membuat:

1. Gurita cuci bersih, buang kotorannya. Siram garam, gosok^2 buang lender yg. dikulitnya, cuci bersih potong pendek aduk bumbu pengasin. Rebus dalam air mendidth. Angkat.
2. Daging kurus cuci bersih, rebus dalam air mendidih, angkat. Potong kotak^2.
3. Kuilt jeruk rendam pakai air dingin sampai empuk. Kerok seratnya. Bahan obat taruh kedalam bahan obat tas.
4. Panci besar kasih air, semua tarah ke dalam. Pakai api besar masak sampai mendidih, pakai api sedang masak 60 menit, lalu kasih garam. Jadilah.

椰子紅豆通草鮮八爪魚湯

Double-steamed Soup with Cordyceps and Black-skinned Chicken
Kuah Cordyceps dan Ayam Berkulit Hitam

- one month after giving birth
- take it once or twice a week
- do not take the soup if you have a disease caused by external factors

Ingredients:

19 g cordyceps
4 black dates
38 g Huai Shan
1/2 black-skinned chicken
2 ginger slices
1000 ml cold boiling water

Method:

1. Rinse, skin and gut the black-skinned chicken. Cut off the tip of the chicken wings, head, neck and buttocks with scissors. Rinse, blanch in boiling water and drain. Cut into medium-sized pieces. Set aside.
2. Rinse the cordyceps, black dates and Huai Shan.
3. Put all the ingredients into a ceramic container for double-steaming. Put the lid on and double-steam over low heat for 4 hours. Season with salt. Serve.

- satu bulan setelah melahirkan
- makan sekali atau dua kali seminggu
- jika penyakit yang disebabkan oleh faktor eksternal, tidak makan

Bahan:

19 g cordyceps
4 bj. Nanzao (jujube)
38 g Huai Shan
1/2 ek. ayam berkulit hitam
2 lb. jahe
1000 ml air mendidih dingin

冬蟲夏草烏雞燉湯

Cara Membuat:

1. Ayam berkulit hitam cuci bersih, kupas kulit, buang kotoran dalamnya. Buang ujung sayap ayam, kepala, leher, pantat. Cuci bersih, rebus dalam air mendidih, angkat, tiriskan, potong kotak^2.
2. Cordyceps, Nanzao, Huai Shan cuci bersih.
3. Masukkan semua bahan ke dalam wadah keramik untuk double-mengukus. Tutup dengan tutupnya, pakai api sedang masak 4 jam, kasih garam. Jadilah.

Soup with Qi Zi, Dried Longan, Sang Ji Sheng and Lotus Root
Kuah Qi Zi, Lengkeng Kering, Sang Ji Sheng dan Akar Teratai

- 12 days after giving birth
- take it once or twice a week
- do not take the soup if you have a disease caused by external factors

Ingredients:

19 g Qi Zi
15 g dried longan aril
38 g Sang Ji Sheng
4 organic honey dates
900 g lotus root
1.2 kg pork shoulder bone
2 ginger slices
3000 ml water

Method:

1. Rinse the pork shoulder bone. Blanch in boiling water to remove grease.
2. Skin the lotus root, rinse and cut into medium-sized pieces. Set aside.
3. Rinse the Qi Zi, dried longan aril and Sang Ji Sheng. Set aside. Pack the herbal ingredients in a paper bag for decoction or in a muslin bag.
4. Pour water into a tall stock pot and add all the ingredients. Bring to the boil over high heat, turn to low-medium heat and cook for 90 minutes. Season with salt. Serve.

- -

- setelah 12 hari setelah melahirkan
- makan sekali atau dua kali seminggu
- jika penyakit yang disebabkan oleh faktor eksternal, tidak makan

Bahan:

19 g Qi Zi
15 g daging lengkeng kering
38 g Sang Ji Sheng
4 bj. organik korma
900 g akar teratai
1.2 kg tulang iga
2 lb. jahe
3000 ml air

Cara Membuat:

1. Tulang iga cuci bersih, rebus dalam air mendidih, angkat, buang gajihnya.
2. Akar teratai kupas kulit, cuci bersih, potong kotak[2].
3. Qi Zi, daging lengkeng kering, Sang Ji Sheng cuci bersih. Bahan obat taruh kedalam bahan obat tas.
4. Panci besar kasih air semua bahan tutup panci. Pakai api besar masak sampai mendidih, pakai api sedang masak 90 menit, lalu kasih garam. Jadilah.

杞子桂圓桑寄生蓮藕湯

Fish Maw Soup with Bei Qi, Dang Shen and Hai Yu Zhu
Kuah Bei Qi, Dang Shen, Hai Yu Zhu dan Ikan Perut

- 12 days after giving birth
- take it once or twice a week
- do not take the soup if you have a disease caused by external factors

Ingredients:

15 g roasted Bei Qi
19 g Dang Shen
150 g frozen whole Hai Yu Zhu
75 g dried fish maw
300 g lean pork
4 organic honey dates
1 small piece dried tangerine peel
2 ginger slices
3000 ml water

Ingredients for rehydrating fish maw:

6 ginger slices
2 sprigs spring onion
1/2 bowl Shaoxing wine
2 large pots water

Method:

1. Soak the dried fish maw in water overnight. Change the water and soak again for 8 hours. Cut open with scissors and remove the filth inside. Bring a pot of water to the boil and put in the fish maw. Turn off heat and leave it covered until the water cools. Take out the fish maw and cut into several pieces with scissors. Bring another pot of water to the boil with ginger and spring onion. Put in the fish maw and boil for 3 minutes. Sprinkle with the wine and boil for 2 minutes. Turn off heat and leave it covered until the water cools. Take out the fish maw and rinse until it is completely clean.
2. Rinse, skin and slice the Hai Yu Zhu. Set aside. Blanch the lean pork in boiling water and cut into pieces.
3. Soak the Bei Qi and Dang Shen in water for 2 hours. Discard the water.
4. Soak the dried tangerine peel in water until soft and scrape off the pith.
5. Put all the ingredients into a pot. Add water and bring to the boil over high heat. Turn down the heat and simmer for 3 hours. Season with salt. Serve.

- setelah 12 hari setelah melahirkan
- makan sekali atau dua kali seminggu
- jika penyakit yang disebabkan oleh faktor eksternal, tidak makan

Bahan:

15 g panggang Bei Qi
19 g Dang Shen
150 g beku seluruh Hai Yu Zhu
75 g kering ikan perut
300 g daging kurus
4 bj. organik korma
pojokan kulit jeruk kering
2 lb. jahe
3000 ml air

Bahan untuk kering ikan perut rehydrating:

6 lb. jahe
2 tangkai daun bawang
1/2 bowl arak masak
2 panci besar air

Cara Membuat:

1. Kering ikan perut rendam dalam air semalam. Ganti air lalu rendam lagi 8 jam. Potong buang kotoran dalamnya. Taruh panci berisi air rebus mendidih taruh ikan perut, matikan api, air mendidih ambil ikan perut, potong kotak², taruh panci rebus air, taruh jahe, daun berambang, ikan perut, masak 3 menit, kasih arak masak 2 menit, matikan api tunggu air dinggin ambil ikan perut.
2. Hai Yu Zhu cuci bersih, kupas kulit, potong kotak². Daging kurus rebus dalam air mendidih, angkat, potong kotak².
3. Bei Qi and Dang Shen pakai air rendam 2 jam, buang airnya.
4. Kuilt jeruk rendam pakai air dingin sampai empuk. Kerok seratnya.
5. Semua bahan tutup panci. Pakai api besar masak sampai mendidih, pakai api sedang masak 3 jam, lalu kasih garam. Jadilah.

北芪黨參海玉竹花膠湯

Pork Shin Soup with Shi Hu, Nu Zhen Zi, Tien Dong, Bei Sha Shen and Nan Qi
Kuah Shi Hu, Nu Zhen Zi, Tien Dong, Bei Sha Shen, Nan Qi dan Kaki Babi

- within 12 days after giving birth
- substitute the ginger for dried tangerine peel for women after caesarean
- take it twice or three times a week
- do not take the soup if you have a disease caused by external factors

Ingredients:

15 g Huo Shan Shi Hu
19 g Nu Zhen Zi
38 g Tien Dong
38 g Bei Sha Shen
38 g Nan Qi
600 g pork shin
4 organic honey dates
2 ginger slices
2000 ml water

Method:

1. Rinse the pork shin and blanch in boiling water. Rinse again and drain. Set aside.
2. Rinse the Huo Shan Shi Hu, Nu Zhen Zi, Tien Dong, Bei Sha Shen and Nan Qi. Drain. Soak the Tien Dong and Bei Sha Shen in water for 2 hours. Discard the water. Pack the herbal ingredients in a paper bag for decoction or in a muslin bag.
3. Put all the ingredients and water into a pot. Bring to the boil over high heat, turn down the heat and simmer for 2 hours. Season with salt. Serve.

- dalam waktu 12 hari setelah melahirkan
- wanita setelah operasi caesar, pengganti jahe untuk kulit jeruk kering
- makan dua kali atau tiga kali seminggu
- jika penyakit yang disebabkan oleh faktor eksternal, tidak makan

Bahan:

15 g Huo Shan Shi Hu
19 g Nu Zhen Zi
38 g Tien Dong
38 g Bei Sha Shen
38 g Nan Qi
600 g kaki babi
4 bj. organik korma
2 lb. jahe
2000 ml air

Cara Membuat:

1. Kaki babi cuci bersih, rebus dalam air mendidih, angkat, tiriskan.
2. Huo Shan Shi Hu, Nu Zhen Zi, Tien Dong, Bei Sha Shen, Nan Qi cuci bersih, tiriskan. Tien Dong, Bei Sha Shen pakai air rendam 2 jam, buang airnya. Bahan obat taruh kedalam bahan obat tas.
3. Panci besar kasih air semua bahan tutup panci. Pakai api besar masak sampai mendidih, pakai api sedang masak 2 jam, lalu kasih garam. Jadilah.

石斛女貞子天冬北沙參南芪豬腱湯

Beef Shank Soup with Tian Qi, Prepared Di Huang, Red Dates and Lily Bulbs
Kuah Tian Qi, Siap Di Huang, Tanggal, Lily Bulbs dan Kaki Sapi

- 12 days after giving birth
- take it once or twice a week

- do not take the soup if you have abnormally profuse lochia rubra

Ingredients:

11 g prepared Tian Qi
19 g prepared Di Huang
10 red dates (pitted)
38 g dried lily bulbs
600 g beef shank
1 small piece dried tangerine peel
4 organic honey dates
2 ginger slices
2500 ml water

Method:

1. Rinse the beef shank. Bring water to the boil, turn off heat, add the beef shank and blanch for 1 minute. Take out and rinse again. Set aside.
2. Rinse the prepared Tian Qi, prepared Di Huang and red dates. Soak the dried lily bulbs in water for 2 hours. Discard the water. Pack the herbal ingredients in a paper bag for decoction or in a muslin bag.
3. Put all the ingredients and water into a pot. Bring to the boil over high heat, turn down the heat and simmer for 2 hours. Season with salt. Serve.

- setelah 12 hari setelah melahirkan
- makan sekali atau dua kali seminggu
- wanita dengan berlimpah lokia rubra, tidak makan

Bahan:

11 g siap Tian Qi
19 g siap Di Huang
10 bj. tanggal (buang bijinya)
38 g kering lily bulbs
600 g kaki sapi
pojokan kulit jeruk kering
4 bj. organik korma
2 lb. jahe
2500 ml air

Cara Membuat:

1. Kaki sapi cuci berish. Masak air mendidih, matikan, masukkan kedalam, masak 1 menit, ambil keluar cuci bersih.
2. Tian Qi, Di Huang dan tanggal cuci bersih. Lily bulbs pakai air rendam 2 jam. Buang airnya. Bahan obat taruh kedalam bahan obat tas.
3. Panci besar kasih air semua bahan tutup panci. Pakai api besar masak sampai mendidih, pakai api sedang masak 2 jam. Lalu kasih garam. Jadilah.

田七熟地紅棗百合牛腱湯

Frog Soup with Du Zhong and Fox Nuts
Kuah Du Zhong, Fox Nuts dan Katak

- 12 days after giving birth
- take it once or twice a week

- do not take the soup if you have a disease caused by external factors

Ingredients:

15 g organic dried daylily flower
38 g dried wood ear fungus
15 g Du Zhong
(processed with salted water)
38 g fox nuts
(processed with salted water)
300 g Chinese edible frog
3 ginger slices
5 red dates (pitted)
2 tbsp glutinous rice wine
1000 ml water

Method:

1. Skin the frog and remove its backbone, fingers and toes. Rinse and blanch in boiling water. Drain and set aside.
2. Heat oil in a wok. Sauté the ginger until fragrant. Add the frog and stir-fry for 2-3 minutes to remove its muddy smell.
3. Soak the dried wood ear fungus in water until soft. Remove the stalks, rinse and tear to bite-sized pieces. Rinse the dried daylily flower.
4. Pack the herbal ingredients in a paper bag for decoction or in a muslin bag.
5. Put all the ingredients and water into a pot. Bring to the boil over high heat, turn down the heat and simmer for 20 minutes. Add the wine and continue to cook for 20 minutes. Season with salt. Serve.

- setelah 12 hari setelah melahirkan
- makan sekali atau dua kali seminggu
- jika penyakit yang disebabkan oleh faktor eksternal, tidak makan

Bahan:

15 g organik kering bunga bakung
38 g kering jamur
15 g Du Zhong (diproses dalam air asin)
38 g fox nuts (diproses dalam air asin)
300 g katak
3 lb. jahe
5 bj. tanggal (buang bijinya)
2 sdk. arak beras
1000 ml air

Cara Membuat:

1. Kupas kulit katak, buang jari², kerongkongan, cuci bersih, rebus dalam air mendidih, angkat, tiriskan.
2. Panaskan wajan dengan minyak, taruh jahe, katak, tumis 2-3 menit.
3. Kering jamur pakai air sampai empuk, petik yg., cuci bersih, sobek menjadi potongan kecil. Bunga bakung cuci bersih.
4. Bahan obat taruh kedalam bahan obat tas.
5. Panci besar kasih air semua bahan tutup panci. Pakai api besar masak sampai mendidih, pakai api sedang masak 20 menit, kasih arak masak 20 menit, lalu kasih garam. Jadilah.

杜仲肇實田雞湯

Sea Cucumber and Abalone Soup with Shiitake Mushrooms
Kuah Timun Laut, Pauhi dan Jamur Wangi Kering

- within 12 days after giving birth
- skip the ginger for women after caesarean
- can be taken every day

- do not take the soup if you have a disease caused by external factors

Ingredients:

10 high quality dried shiitake mushrooms
900 g frozen sea cucumber
600 g frozen Australian abalone
300 g lean pork
150 g peanuts (shelled, with skin)
1 small piece dried tangerine peel
3 ginger slices
2500 ml water

Method:

1. Rinse all the ingredients. Soak the dried shiitake mushrooms in water until soft and then rinse. Separate the mushroom body and stalks. For meaty mushrooms, cut into half with scissors to release flavour. Blanch the lean pork in boiling water to remove grease and cut into chunks. Remove the internal organs of the abalone, rinse and coarsely slice.
2. Soak the dried tangerine peel in water until soft and scrape off the pith.
3. Defrost the sea cucumber, cut open with scissors, remove the internal organs and filth and then rinse. Blanch in boiling water with Shaoxing wine for 2 minutes.
4. Put all the ingredients and water into a pot. Bring to the boil over high heat, turn down the heat and simmer for 2 hours. Season with salt. Serve.

- dalam waktu 12 hari setelah melahirkan
- melewatkan jahe untuk wanita setelah operasi caesar
- dapat makan setiap hari
- jika penyakit yang disebabkan oleh faktor eksternal, tidak makan

Bahan:

10 bj. jamur wangi kering
900 g timun laut beku
600 g Australia pauhi beku
300 g daging kurus
150 g kacang tanah (berkerang, dengan kulit)
pojokan kulit jeruk kering
3 lb. jahe
2500 ml air

冬菇海參鮑魚湯

Cara Membuat:

1. Semua bahan cuci bersih. Jamur wangi kering pakai air sampai empuk, cuci bersih, jamur kering wangi daging sama batangnya di iris pisah. Daging kurus rebus dalam air mendidih, angkat, potong kotak². Pauhi buang kotoran dalamnya, cuci bersih, potong kotak².
2. Kuilt jeruk rendam pakai air dingin sampai empuk. Kerok seratnya.
3. Timun laut beku di keluarkan sampai tidok ber-es, di potong, buang kotoran di dalamnya, kasih arak, rebus dalam air mendidih 2 menit, angkat.
4. Panci besar kasih air semua bahan tutup panci. Pakai api besar masak sampai mendidih, pakai api sedang masak 2 jam, lalu kasih garam. Jadilah.

Tilapia Soup with Black Beans, Chuan Xiong and Tian Ma
Kuah Kacang Hitam, Chuan Xiong, Tian Ma dan Ikan Tambang

- 12 days after giving birth
- take it once every two days
- women having abnormally profuse lochia rubra should be cautious of taking it

Ingredients:

75 g black beans
75 g dried small wild cloud ear fungus
11 g Chuan Xiong
4 jujubes
15 g Tian Ma
 (processed with ginger juice)
11 g Bai Zhi
1.2 kg tilapia
4 ginger slices
2300 ml water

Method:

1. Scale, gut and gill the tilapia. Rinse and wipe dry. Fry the fish with a little oil until both sides are light brown. Remove and wipe the oil away the fish with kitchen paper. Set aside.
2. Soak the small cloud ear fungus in water until soft. Remove the stalks and rinse.
3. Pack the herbal ingredients in a paper bag for decoction or in a muslin bag.
4. Put water and all the ingredients into a pot. Bring to the boil over high heat, turn down the heat and simmer for 2 hours. Season with salt. Serve.

- setelah 12 hari setelah melahirkan
- makan sekali setiap dua hari
- wanita dengan berlimpah lokia rubra, berhati-hati

Bahan:

75 g kacang hitam
75 g kering kecil liar jamur kuping
11 g Chuan Xiong
4 bj. jujubes
15 g Tian Ma (diproses dalam sari jahe)
11 g Bai Zhi
1.2 kg tambang
4 lb. jahe
2300 ml air

Cara Membuat:

1. Tambang di bersihkan sisiknya, buang isang dan kotoran di dalamanya cuci bersih. Tambang goreng sampai agak kuning. Pakai kertas dapur menyerap minyaknya.
2. Jamur kuping pakai air sampai empuk, petik yg., cuci bersih.
3. Bahan obat taruh kedalam bahan obat tas.
4. Panci besar kasih air semua bahan tutup panci. Pakai api besar masak sampai mendidih, pakai api sedang masak 2 jam, lalu kasih garam. Jadilah.

黑豆川芎天麻鯛魚湯

Salmon Head Soup with Lotus Seeds, Red Beans and Tong Cao
Kuah Biji Teratai, Kacang Merah, Tong Cao dan Salmon Kepala

- within 12 days after giving birth
- substitute the ginger for dried tangerine peel for women after caesarean
- can be taken every day
- skip the chicken if you have a disease caused by external factors

Ingredients:

75 g lotus seeds (with skin)
75 g red beans
600 g salmon head
1 chicken (about 900 g)
11 g Tong Cao (ricepaper pith)
3 ginger slices
2200 ml water

Method:

1. Rinse the salmon head and cut into pieces. Fry with a little oil until fragrant. Wipe the oil off and set aside.
2. Rinse, skin and gut the chicken. Cut off the tip of the chicken wings, head, neck and buttocks with scissors. Rinse and blanch in boiling water to remove grease. Set aside.
3. Rinse the lotus seeds and set aside.
4. Put all the ingredients into a pot. Add water, bring to the boil over high heat, turn down the heat and simmer for 2 hours. Season with salt. Serve.

- dalam waktu 12 hari setelah melahirkan
- wanita setelah operasi caesar, pengganti jahe untuk kulit jeruk kering
- dapat makan setiap hari
- jika penyakit yang disebabkan oleh faktor eksternal, melewatkan ayam

Bahan:

75 g biji teratai (dengan kulit)
75 g kacang merah
600 g salmon kepala
1 ek. ayam (900 g)
11 g Tong Cao
3 lb. jahe
2200 ml air

蓮子紅豆通草三文魚頭湯

Cara Membuat:

1. Salmon kepala cuci bersih, potong kotak^2, goreng pakai minyak sedikit.
2. Ayam cuci bersih, kupas kulitnya, buang kotoran dalamnya, buang pucuk sewiwi ayam, leher, pantat. Cuci bersih, rebus dalam air mendidih, angkat.
3. Biji teratai cuci bersih.
4. Panci besar kasih air semua bahan tutup panci. Pakai api besar masak sampai mendidih, pakai api sedang masak 2 jam, lalu kasih garam. Jadilah.

Swamp Eel and Peanut Soup
Kuah Belut dan Kacang Tanah

- within 12 days after giving birth
- substitute the ginger for dried tangerine peel for women after caesarean
- take it once or twice a week
- do not take the soup if you have a disease caused by external factors

Ingredients:

600 g swamp eel
150 g peanuts (with skin)
5 red dates (pitted)
11 g Wang Bu Liu Xing Zi
 (vaccaria seed)
150 g lean pork
2200 ml water
3 ginger slices

Seasoning:

1 tbsp rice wine
1/2 tsp ground white pepper
1/3 tsp salt

Method:

1. Rinse the lean pork, blanch in boiling water and cut into pieces. Set aside.
2. Gut the swamp eel and rinse. Rub the skin and the inside with coarse salt to remove mucus. Rinse again and cut into pieces. Set aside. Heat oil in a wok. Sauté the ginger and swamp eel until aromatic. Wipe the oil off the eel with kitchen paper. Set aside.
3. Rinse the peanuts, red dates and Wang Bu Liu Xing Zi. Set aside. Pack the Wang Bu Liu Xing Zi in a paper bag for decoction.
4. Put all the ingredients into a pot. Add water, bring to the boil over high heat, turn down the heat and simmer for 1 hour. Add the seasoning and give a good stir. Serve.

- -

- dalam waktu 12 hari setelah melahirkan
- wanita setelah operasi caesar, pengganti jahe untuk kulit jeruk kering
- makan sekali atau dua kali seminggu
- jika penyakit yang disebabkan oleh faktor eksternal, tidak makan

Bahan:

600 g belut
150 g kacang tanah (dengan kulit)
5 bj. tanggal (buang bijinya)
11 g Wang Bu Liu Xing Zi
150 g daging kurus
2200 ml air
3 lb. jahe

Bumbu:

1 sdk. beras anggur
1/2 sdt. merica
1/3 sdt. garam

黃
鱔
花
生
湯

Cara Membuat:

1. Daging kurus, cuci bersih, rebus dalam air mendidih, angkat, potong kotak^2.
2. Belut buang kotoran dalamnya, cuci bersih, siram garam, gosok^2 buang lender yg. Dikulitnya, cuci bersih, potong kotak^2. Panaskan kewali dng, tumis jahe dan belut. Pakai kertas dapur mengelap minyaknya.
3. Kacang tanah, tanggal, Wang Bu Liu Xing Zi cuci bersih. Bahan obat taruh kedalam bahan obat tas.
4. Panci besar kasih air semua bahan tutup panci. Pakai api besar masak sampai mendidih, pakai api sedang masak 1 jam, lalu kasih bumbu. Jadilah.

Conch Soup with Walnuts, Snow Fungus and Dried Figs
Kuah Kenari, Jamur Putih Kering, Kering Ara dan Kulit Kerang

- within 12 days after giving birth
- can be taken every day
- skip the ginger for women after caesarean
- do not take the soup if you have a disease caused by external factors

Ingredients:

75 g walnuts (with skin)
75 g dried snow fungus
6 organic sun-dried figs
500 g frozen shelled Australian conch
8 fresh chicken feet (medium-sized)
1 small piece dried tangerine peel or 4 ginger slices
2800 ml water

Method:

1. Soak the dried snow fungus in water until soft. Remove the stalks and impurities and tear into several pieces.
2. Soak the dried figs in water until soft. Cut in half with scissors.
3. Defrost the shelled conch, blanch in boiling water and cut into pieces. Set aside.
4. Rinse the chicken feet, blanch in boiling water, rinse again and drain. Set aside.
5. Soak the dried tangerine peel in water until soft and scrape off the pith.
6. Put all the ingredients into a pot. Add water, bring to the boil over high heat, turn down the heat and simmer for 1.5 hours. Season with salt. Serve.

- dalam waktu 12 hari setelah melahirkan
- melewatkan jahe untuk wanita setelah operasi caesar
- dapat makan setiap hari
- jika penyakit yang disebabkan oleh faktor eksternal, tidak makan

Bahan:

75 g kenari (dengan kulit)
75 g jamur putih kering
6 bj. organik kering ara
500 g Australia kulit kerang beku
8 ek. segar kaki ayam (sedang)
pojokan kulit jeruk kering atau 4 lb. jahe
2800 ml air

核桃雪耳無花果響螺湯

Cara Membuat:

1. Jamur putih kering pakai air sampai empuk, petik yg., cuci bersih. Sobek menjadi potongan kecil.
2. Kering ara pakai air sampai empuk, potong 2 bagian.
3. Kulit kerang laut beku di keluarkan sampai tidak ber-es, rebus dalam air mendidih, angkat, potong kotak^2.
4. Kaki ayam cuci bersih, rebus dalam air mendidih, angkat, tiriskan.
5. Kulit jeruk rendam pakai air dingin sampai empuk. Kerok seratnya.
6. Panci besar kasih air semua bahan tutup panci. Pakai api besar masak sampai mendidih, pakai api sedang masak 1.5 jam lalu kasih garam. Jadilah.

Tomato Chicken Soup with Soy Bean Sprouts and Haw
Kuah Kedelai Kecambah, Pagar Keliling, Tomat dan Ayam

- within 12 days after giving birth
- also suitable for women after caesarean
- can be taken every day
- substitute the chicken for pork if you have a disease caused by external factors

Ingredients:

375 g tomatoes
300 g chicken breast
150 g soy bean sprouts
19 g haw
75 g leeks
900 ml water
salt
1 tsp sugar

Method:

1. Rinse the chicken breast, remove the fascia (thin membrane) and cut into pieces. Set aside.
2. Rinse the rest ingredients. Cut the tomatoes into medium-sized pieces. Pack the haw in a paper bag for decoction.
3. Remove the root of the leeks and soy bean sprouts. Finely shred the leeks.
4. Put the haw, tomatoes, chicken breast and water into a pot. Bring to the boil over high heat, turn to medium heat and cook for 20 minutes. Add the soy bean sprouts and cook until done. Put in the leeks, salt and sugar and cook for a moment. Serve.

- dalam waktu 12 hari setelah melahirkan
- wanita setelah operasi caesar bisa makan
- dapat makan setiap hari
- jika penyakit yang disebabkan oleh faktor eksternal, pengganti ayam untuk daging babi

Bahan:

375 g tomat
300 g dada ayam
150 g kedelai kecambah
19 g pagar keliling
75 g daun bawang
900 ml air
sedikit garam
1 sdt. gula

Cara Membuat:

1. Dada ayam, cuci bersih, buang jalur, potong kotak².
2. Semua bahan cuci bersih. Tomat potong kotak². Bahan obat taruh kedalam bahan obat tas.
3. Kedelai kecambah dan daun bawang buang akarnya, daun bawang potong panjang tipis.
4. Panci besar kasih air tomat, dada ayam, pager keliling tutup panci. Pakai api besar masak sampai mendidih, pakai api sedang masak 20 menit, kedelai kecambah masak sampai matang. Daun bawang, garam, gula masak sebentar. Jadilah.

番茄雞肉大豆芽山楂湯

Vegetarian Soup with Fresh and Dried Lily Bulbs and Pumpkin
Vegetarian Kuah Bunga Bakung Segar dan Kering, dan Labu Kuning

- within 12 days after giving birth
- a vegetarian diet
- can be taken every day
- substitute the ginger for dried tangerine peel for women after caesarean

Ingredients:

4 fresh lily bulbs
38 g dried lily bulbs
75 g shelled chestnuts
150 g black beans
900 g skinned pumpkin
 (cut into pieces)
2 ginger slices
2000 ml water

Method:

1. Soak the dried lily bulb, shelled chestnuts and black beans in water for about 1 hour. Discard the water.
2. Rinse the fresh lily bulbs and skinned pumpkin. Set aside.
3. Put all the ingredients (except the fresh lily bulbs) into a pot. Add water, bring to the boil over high heat, turn down the heat and simmer for 1 hour. Put in the fresh lily bulbs and cook for 10 minutes. Season with salt. Serve.

- dalam waktu 12 hari setelah melahirkan
- wanita setelah operasi caesar, pengganti jahe untuk kulit jeruk kering
- vegetarian bisa makan
- dapat makan setiap hari

金銀百合南瓜素湯

Bahan:

4 bj. segar bunga bakung
38 g kering bunga bakung
75 g daging sarangan
150 g kacang hitam
900 g labu kuning
(potong kotak^2)
2 lb. jahe
2000 ml air

Cara Membuat:

1. Kering bunga bakung, daging sarangan dan kacang hitam pakai air 1 jam. Buang air.
2. Segar bunga bakung dan labu kuning, cuci bersih.
3. Panci besar kasih air semua bahan tutup panci. Pakai api besar masak sampai mendidih, pakai api sedang masak 1 jam, segar bunga bakung masar 10 menit, lalu kasih garam. Jadilah.

Vegetarian Soup with Elm Fungus and Yang Huo Ye
Vegetarian Kuah Jamur Elm dan Yang Huo Ye

- 12 days after giving birth
- can be taken every day
- a vegetarian diet
- take less if you suffer from excessive internal heat

Ingredients:

75 g dried elm fungus
38 g dried shiitake mushrooms
600 g beet
19 g Qi Zi
75 g Huai Shan
300 g shelled coconut
15 g Yang Huo Ye
3 ginger slices
2300 ml water

Method:

1. Rinse all the ingredients. Soak the dried elm fungus and shiitake mushrooms in water until soft. Rinse and separate the mushroom body and stalks. Remove the stalks of the elm fungus.
2. Peel the beet, rinse and cut into pieces. Cut the shelled coconut into pieces.
3. Pack the Yang Huo Ye in a paper bag for decoction.
4. Put all the ingredients into a pot. Add water, bring to the boil over high heat, turn down the heat and simmer for 1.5 hours. Season with salt. Serve.

- setelah 12 hari setelah melahirkan
- vegetarian bisa makan
- dapat makan setiap hari
- jika penyakit panas internal yang berlebihan, makan sedikit

Bahan:

75 g jamur elm kering
38 g jamur wangi kering
600 g ubi bit
19 g Qi Zi
75 g Huai Shan
300 g daging kelapa
15 g Yang Huo Ye
3 lb. jahe
2300 ml air

乾榆耳羊藿葉素湯

Cara Membuat:

1. Semua bahan cuci bersih. Jamur elm kering dan jamur wangi kering cuci bersih, rendam dalam air sampai empuk, buang tulangnya.
2. Bit cuci bersih, kupas kulit, potong kotak^2. Daging kelapa potong kotak^2.
3. Bahan obat taruh kedalam bahan obat tas.
4. Panci besar kasih air semua bahan tutup panci. Pakai api besar masak sampai mendidih, pakai api sedang masak 1.5 jam, lalu kasih garam. Jadilah.

Double-steamed Chicken Soup with Deer Antler and Ginseng
Kuah Rusa Tanduk, Ginseng dan Ayam Tua

- one month after giving birth
- take it once a week

- do not take the soup if you suffer from excessive internal heat or have a disease caused by external factors

Ingredients:

4 g deer antler
6 g ginseng
1/2 old chicken
2 ginger slices
3 bowls cold boiling water

Method:

1. Rinse, skin and gut the chicken. Cut off the tip of the chicken wings, head, neck and buttocks with scissors. Rinse, blanch in boiling water and drain. Cut into medium-sized pieces. Set aside.
2. Put all the ingredients into a ceramic container for double-steaming. Put the lid on and double-steam over low heat for 4 hours. Season with salt. Serve.

- satu bulan setelah melahirkan
- makan sekali seminggu
- jika penyakit yang disebabkan oleh faktor eksternal
- atau panas internal yang berlebihan, tidak makan

Bahan:

4 g rusa tanduk
6 g ginseng
1/2 ek. ayam tua
2 lb. jahe tua
3 bowls dingin air mendidih

鹿茸人參燉老雞

Cara Membuat:

1. Ayam tua cuci bersih, kupas kulitnya, buang kotoran dalamnya, buang pucuk sewiwi ayam, leher, pantat. Cuci bersih, rebus dalam air mendidih, angkat, tiriskan, potong kotak^2.
2. Masukkan semua bahan ke dalam wadah keramik untuk double-mengukus. Tutup dengan tutupnya, pakai api sedang, masak 4 jam, kasih garam. Jadilah.

Pearl Mussel, Shi Hu and Sea Cucumber Soup

Kuah Pearl Remis, Shi Hu dan Timun Laut

- within 12 days after giving birth
- substitute the ginger for dried tangerine peel for women after caesarean
- can be taken every day
- do not take the soup if you have a disease caused by external factors

Ingredients:

38 g shelled pearl mussel
11 g Huo Shan Shi Hu
300 g frozen sea cucumber
600 g lean pork
3 ginger slices

Method:

1. Blanch the lean pork to remove grease and cut into pieces. Soak the shelled pearl mussel in water for half an hour. Rinse and cut in half with scissors.
2. Defrost the sea cucumber, cut open with scissors, remove the internal organs and filth and then rinse. Blanch in boiling water with Shaoxing wine for 2 minutes.
3. Put all the ingredients and water into a pot. Bring to the boil over high heat, turn down the heat and simmer for 2 hours. Season with salt. Serve.

- -

- dalam waktu 12 hari setelah melahirkan
- wanita setelah operasi caesar, pengganti jahe untuk kulit jeruk kering
- dapat makan setiap hari
- jika penyakit yang disebabkan oleh faktor eksternal, tidak makan

Bahan:

38 g daging pearl remis
11 g Huo Shan Shi Hu
300 g timun laut beku
600 g daging kurus
3 lb. jahe

珍珠肉石斛海參湯

Cara Membuat:

1. Daging kurus rebus dalam air mendidih, angkat, potong kotak^2. Daging pearl remis pakai air rendam 30 menit, cuci bersih, potong 2 bargian.
2. Timun laut di keluarkan sampai tidak ber-es, potong buang kotoran dalamnya, cuci bersih, kasih arak, rebus dalam air mendidih 2 menit, angkat.
3. Panci besar kasih air semua bahan tutup panci. Pakai api besar masak sampai mendidih, pakai api sedang masak 2 jam, lalu kasih garam. Jadilah.

Soup with Du Zhong, Niu Da Li and Jin Gou Ji
Kuah Du Zhong, Niu Da Li dan Jin Gou Ji

• can be taken once or twice a week

Ingredients:

19 g Du Zhong
15 g Niu Da Li
11 g Jin Gou Ji
6 honey dates
1 small piece dried tangerine peel
600-900 g pork tail bone
2500 ml water

Method:

1. Pack the herbal ingredients in a paper bag for decoction or in a muslin bag.
2. Soak the dried tangerine peel in water until soft and scrape off the pith.
3. Rinse the pork tail bone, blanch in boiling water to remove grease and cut into pieces.
4. Put all the ingredients and water into a pot. Bring to the boil over high heat, turn down the heat and simmer for 2 hours. Season with salt. Serve.

- -

• makan sekali atau dua kali seminggu

Bahan:

19 g Du Zhong
15 g Niu Da Li
11 g Jin Gou Ji
6 bj. korma
pojokan kulit jeruk kering
600-900 g tulang babi ekor
2500 ml air

Cara Membuat:

1. Bahan obat taruh kedalam bahan obat tas.
2. Kulit jeruk rendam pakai air dingin sampai empuk. Kerok seratnya.
3. Tulang babi ekor, cuci bersih, rebus dalam air mendidih, angkat, potong kotak².
4. Panci besar kasih air semua bahan tutup panci. Pakai api besar masak sampai mendidih, pakai api sedang masak 2 jam, lalu kasih garam. Jadilah.

杜仲牛大力金狗脊湯

Five Black Treasure Soup
Kuah Lima Bahan Mahal Hitam

- skip the ginger for women after caesarean
- can be taken once or twice a week

Ingredients:

11 g processed He Shou Wu
75 g black beans
38 g black wood ear fungus
4 black dates
1 black-skinned chicken
3 ginger slices
2200 ml water

Method:

1. Rinse the black wood ear fungus, remove the stalks and tear to bite-sized pieces.
2. Rinse, skin and gut the black-skinned chicken. Remove the head, neck, the tip of the chicken wings and buttocks. Rinse and blanch in boiling water.
3. Put all the ingredients and water into a pot. Bring to the boil over high heat, turn down the heat and simmer for 2 hours. Season with salt. Serve.

- -

- melewatkan jahe untuk wanita setelah operasi caesar
- makan sekali atau dua kali seminggu

Bahan:

11 g siap He Shou Wu
75 g kacang hitam
38 g jamur hitam
4 bj. memangkas
1 ek. ayam berkulit hitam
3 lb. jahe
2200 ml air

黑色五珍湯

Cara Membuat:

1. Jamur hitam cuci bersih, petik yg, sobek menjadi potongan kecil.
2. Ayam berkulit hitam cuci bersih, kupas kulitnya, buang kotoran dalamnya, buang pucuk sewiwi ayam, leher, pantat. Cuci bersih, rebus dalam air mendidih, angkat.
3. Panci besar kasih air semua bahan tutup panci. Pakai api besar masak sampai mendidih, pakai api sedang masak 2 jam, lalu kasih garam. Jadilah.

Double-steamed Lean Pork Soup with Korean Ginseng and Deer Antler
Kuah Korean Ginseng, Rusa Tanduk dan Daging Kurus

• skip the ginger for women after caesarean
• take it once a week

Ingredients:

8 g Korean ginseng
4 g deer antler
1 small piece dried tangerine peel
2 ginger slices
4 red dates (pitted)
300 g lean pork
1000 ml boiling water

Method:

1. Soak the dried tangerine peel in water until soft and scrape off the pith.
2. Rinse the lean pork, blanch in boiling water and cut into pieces.
3. Put all the ingredients into a ceramic container for double-steaming. Put the lid on, covered and double-steam over medium heat for 4 hours. Take out and season with salt. Serve.

• melewatkan jahe untuk wanita setelah operasi caesar
• makan sekali seminggu

Bahan:

8 g Korean ginseng
4 g rusa tanduk
pojokan kulit jeruk kering
2 lb. jahe
4 bj. tanggal (buang bijinya)
300 g daging kurus
1000 ml air mendidih

高麗參鹿茸燉瘦肉

Cara Membuat:

1. Kulit jeruk rendam pakai air dingin sampai empuk. Kerok seratnya.
2. Daging kurus cuci bersih, rebus dalam air mendidih, angkat, potong kotak^2.
3. Masukkan semua bahan ke dalam wadah keramik untuk double-mengukus. Tutup dengan tutupnya, pakai api sedang, masak 4 jam, lalu kasih garam. Jadilah.

Eight Treasure Soup with Abalone and Marine Fish
Kuah Delapan Harta, Pauhi dan Ikan Laut

- skip the ginger for women after caesarean
- take it once every one to two weeks starting from the fourth month after giving birth

Ingredients:

11 g Dang Gui
11 g prepared Di Huang
11 g Yun Ling
11 g Dang Shen
8 g Chuan Xiong
8 g Bai Shao
8 g Bai Zhu
6 g liquorice
5 jujubes
1-2 abalones
600 g any scaled marine fish
3 ginger slices
2800 ml water

Method:

1. Scale and rinse the fish. Fry the fish with a little oil until both sides are light brown. Remove and wipe the oil away the fish with kitchen paper. Set aside.
2. Remove the internal organs of the abalone, rinse and cut into pieces. Set aside.
3. Pack the herbal ingredients in a paper bag for decoction or in a muslin bag.
4. Put water and all the ingredients into a pot. Bring to the boil over high heat, turn down the heat and simmer for 2 hours. Season with salt. Serve.

- -

- melewatkan jahe untuk wanita setelah operasi caesar
- setelah 3 bulan setelah melahirkan, makan sekali satu atau dua minggu

Bahan:

11 g Dang Gui
11 g siap Di Huang
11 g Yun Ling
11 g Dang Shen
8 g Chuan Xiong
8 g Bai Shao
8 g Bai Zhu
6 g liquorice
5 bj. jujube
1-2 ek. pauhi
600 g ikan laut dengan skala
3 lb. jahe
2800 ml air

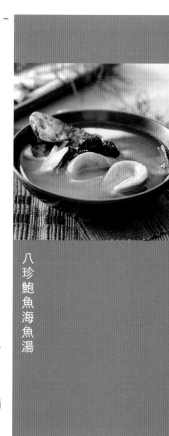

八珍鮑魚海魚湯

Cara Membuat:

1. Ikan bersihkan sisiknya, cuci bersih. Ikan lalu goreng sampai agak kuning. Pakai kertas dapur menyerap minyaknya.
2. Pauhi buang kotoran dalamnya, cuci bersih, potong kotak[2].
3. Bahan obat taruh kedalam bahan obat tas.
4. Panci besar kasih air semua bahan tutup panci. Pakai api besar masak sampai mendidih, pakai api sedang masak 2 jam, lalu kasih garam. Jadilah.

Whole Nourishing Chicken Soup
Kuah Sempurna Bergizi dan Ayam

- also suitable for women after caesarean
- take it once every one to two weeks starting from the fourth month after giving birth

Ingredients:

8 g Korean ginseng
8 g Dang Gui head
4 g cinnamon
11 g prepared Di Huang
11 g Yun Ling
8 g Chuan Xiong
8 g Bai Shao
8 g Bei Qi
8 g Bai Zhu
6 g liquorice
5 jujubes
1 old hen
3 ginger slices
1 small piece tangerine peel
3000 ml water

Method:

1. Soak the dried tangerine peel in water until soft and scrape off the pith.
2. Skin and gut the chicken. Remove the head, neck, the tip of the chicken wings and buttocks. Rinse and blanch in boiling water.
3. Pack the herbal ingredients in a paper bag for decoction or in a muslin bag.
4. Put all the ingredients and water into a pot. Bring to the boil over high heat, turn down the heat and simmer for 2 hours. Season with salt. Serve.

- wanita setelah operasi caesar bisa makan
- setelah 3 bulan setelah melahirkan, makan sekali satu atau dua minggu

Bahan:

8 g Korean ginseng
8 g Dang Gui kepala
4 g cinnamon
11 g siap Di Huang
11 g Yun Ling
8 g Chuan Xiong
8 g Bai Shao
8 g Bei Qi

8 g Bai Zhu
6 g liquorice
5 bj. jujubes
1 ek. ayam betina tua
3 lb. jahe
pojokan kulit jeruk kering
3000 ml air

Cara Membuat:

1. Kulit jeruk rendam pakai air dingin sampai empuk. Kerok seratnya.
2. Ayam cuci bersih, kupas kulitnya, buang kotoran dalamnya, buang pucuk sewiwi ayam, leher, pantat. Cuci bersih, rebus dalam air mendidih, angkat.
3. Bahan obat taruh kedalam bahan obat tas.
4. Panci besar kasih air semua bahan tutup panci. Pakai api besar masak sampai mendidih, pakai api sedang masak 2 jam, lalu kasih garam. Jadilah.

十全大補雞湯

Tortoise Plastron, Deer Sinew and Beef Bone Soup
Kuah Penyu, Rusa Urat Lutut dan Tulang Sapi

- skip the ginger for women after caesarean
- eat the deer sinews too
- take it once a week in the second and third month after giving birth

Ingredients:

1 large tortoise plastron
225 g dried deer sinews
1.2 kg beef bone
8 red dates (pitted)
1 small piece dried tangerine peel
3 ginger slices
600 g lean pork
3000 ml water

Method:

1. Soak the dried tangerine peel in water until soft and scrape off the pith.
2. Soak the deer sinews in water until soft. Rinse, remove the horny nail and impurities and blanch in boiling water. Take out and tear off the fascia (thin membrane). Cut away the rotten meat and blood vessels with scissors and then rinse. Blanch again in boiling water with 3 slices of ginger or 2 stalks of spring onion for a while. Take out and sprinkle with 1/2 bowl of wine while hot. Mix well. Rinse in water to remove the fishy smell and cut into pieces. Set aside.
3. Rinse the beef bone and blanch in boiling water to remove grease.
4. Rinse the lean pork, blanch in boiling water and cut into pieces.
5. Rinse the tortoise plastron.
6. Put all the ingredients and water into a pot. Bring to the boil over high heat, turn down the heat and simmer for 2.5 hours. Season with salt. Serve.

- melewatkan jahe untuk wanita setelah operasi caesar
- setelah 2-3 bulan setelah melahirkan, makan sekali satu seminggu
- makan rusa urat lutut

Bahan:

1 pt. penyu plastron (besar)
225 g kering rusa urat lutut
1.2 kg tulang sapi
8 bj. tanggal (buang bijinya)
pojokan kulit jeruk kering
3 lb. jahe
600 g daging kurus
3000 ml air

Cara Membuat:

1. Kulit jeruk rendam pakai air dingin sampai empuk. Kerok seratnya.
2. Rusa urat lutut pakai air sampai empuk, cuci bersih, potong kuku kaki, rebus dalam air mendidih, angkat, buang jalur, potong daging busuk dan pembuluh darah, cuci bersih. Kasih 3 jahe dan 2 daun bawang, rebus dalam air mendidih, angkat, kasih 1/2 aduk rata, cuci bersih, potong kotak^2.
3. Tulang sapi cuci bersih, rebus dalam air mendidih, angkat.
4. Daging kurus cuci bersih, rebus dalam air mendidih, angkat, potong kotak^2.
5. Tortoise plastron cuci bersih.
6. Panci besar kasih air semua bahan tutup panci. Pakai api besar masak sampai mendidih, pakai api sedang masak 2.5 jam, lalu kasih garam. Jadilah.

龜鹿牛骨湯

Vegetarian Soup for Strengthening the Kidneys, Sinews and Bones
Vegetarian Kuah untuk Memperkuat Ginjal, Otot dan Tulang

- 12 to 100 days after giving birth
- a vegetarian diet
- take it twice or three times a week

- do not take the soup if you have a disease caused by external factors

Ingredients:

11 g Du Zhong
38 g black beans
38 g walnuts
75 g shelled chestnuts
10 red dates (pitted)
5 honey dates
38 g dried snow fungus
1 small piece dried tangerine peel
2500 ml water

Method:

1. Soak the dried snow fungus in water until soft. Rinse, remove the stalks and tear to bite-sized pieces.
2. Soak the dried tangerine peel in water until soft and scrape off the pith.
3. Put all the ingredients into a pot. Add water and bring to the boil over high heat, turn down the heat and simmer for 1.5 hours. Season with salt. Serve.

補腎壯筋骨素湯

- dari 12 sampai 100 hari setelah melahirkan
- vegetarian bisa makan
- makan dua kali atau tiga kali seminggu
- jika penyakit yang disebabkan oleh faktor eksternal, tidak makan

Bahan:

11 g Du Zhong
38 g kacang hitam
38 g buah kenari
75 g daging sarangan
10 bj. tanggal (buang bijinya)
5 bj. korma
38 g kering jamur kuping putih
pojokan kulit jeruk kering
2500 ml air

Cara Membuat:

1. Kering jamur kuping putih pakai air sampai empuk, petik yg., cuci bersih, sobek menjadi potongan kecil.
2. Kulit jeruk rendam pakai air dingin sampai empuk. Kerok seratnya.
3. Panci besar kasih air semua bahan tutup panci. Pakai api besar masak sampai mendidih, pakai api sedang masak 1.5 jam, lalu kasih garam. Jadilah.

Egg Rice with Octopus and Qi Zi
Nasi Telur Gurita dan Qi Zi

- within 12 days after giving birth
- good to be treated as the meal taken between 3 am to 5 am before dawn
- do not take the rice if you have a disease caused by external factors
- can be taken every day
- For women after caesarean, skip the shredded ginger but marinate the octopus with ginger juice to remove its fishy smell. Rinse the octopus before steaming it with rice.

Ingredients:
300 g fresh octopus
1 tbsp Qi Zi
1 egg (beaten)
1/3 cup pearl rice
1 tbsp shredded ginger
1 tsp black sesame seeds
1 tsp white sesame seeds
1/2 tsp oil
spring onion (chopped)

Marinade:
1/3 tsp sugar
1/3 tsp salt
1/2 tsp sesame oil
1/2 tsp light soy sauce
1 tbsp rice wine

Method:
1. Rinse and gut the octopus. Rub the mucus off with coarse salt and blanch in boiling water. Set aside. Dice the octopus, mix with the marinade and leave for 10 minutes. Heat oil in a wok, sauté the ginger until fragrant. Stir-fry the octopus until sweet-scented and half done. Remove and set aside.
2. Rinse the rice and put into a rice cooker to cook until water dries. Add the octopus, Qi Zi and egg and cook until the rice is done, referring to the signal of the rice cooker.
3. Stir-fry the black and white sesame seeds in a dry wok.
4. Sprinkle spring onion and sesame seeds over the rice. Serve.

- dalam waktu 12 hari setelah melahirkan
- Wanita setelah operasi caesar, pakai sari jahe gurita membumbuinya. Sampai bau amis hilang, angkat, cuci bersih.
- dapat makan setiap hari
- Ini nasi sebaiknya di makan jam 3-5 pagi
- jika penyakit yang disebabkan oleh faktor eksternal, tidak makan

Bahan:
300 g segar gurita
1 sdk. Qi Zi
1 bj. telur ayam (kocok rata)
1/3 ukuran cangkir pearl rice
1 sdk. jahe (potong panjang)
1 sdt. wijen hitam
1 sdt. wijen putih
1/2 sdt. minyak
sedikit daun bawang (potong kecil)

Bumbu Pengasin:
1/3 sdt. gula
1/3 sdt. garam
1/2 sdt. minyak wijen
1/2 sdt. kecap putih
1 sdk. arak beras

八爪魚杞子雞蛋飯

Cara Membuat:
1. Gurita buang kotoran dalamnya, cuci bersih, siram garam, gosok² buang lender yg. dikulitnya, rebus dalam air mendidih, angkat. Gurita potong kecil, pakai bumbu pengasin, aduk rata, rendam 10 menit. Panaskan kewali dng. minyak, kasih jahe, gurita masak setengah matang.
2. Taruh beras ke dalam rice cooker, masak hingga airnya habis, masukkan gurita, Qi Zi dan telur, terus masak sampai matang nasinya.
3. Wijen tumis tanpa minyak.
4. Taruh wijen dan daun bawang diatas nasi. Jadilah.

Garlic Salmon Steak Spaghetti
Spageti Ikan Salmon dan Bawang Putih

- within 12 days after giving birth
- can be taken every day
- also suitable for women after caesarean
- do not take it if you are allergic to the ingredients

Ingredients:

1 organic salmon steak (about 300 g)
1 tbsp grated garlic
dried spaghetti (the amount depending on individual appetite)
1 sprig spring onion (chopped)
1000-2000 ml water (referring to the directions on the package of spaghetti)
1/3 tsp salt (for boiling with spaghetti)
1/2 tsp oil

Marinade:

2 tsp lemon juice
1 tsp sesame oil
1/3 tsp salt
ground black pepper

Method:

1. Rinse the salmon steak and wipe dry. Set aside.
2. Mix the marinade with the salmon steak and leave for 10 minutes.
3. Boil the spaghetti according to the directions on the package and to the degree of chewiness you like. Drain and set aside.
4. Heat oil in a wok and fry the salmon steak over high heat for 1 minute. Flip over and fry again for 1 minute. Remove and set aside. Add the garlic in the same wok and sauté until fragrant. Turn to medium heat, put in the spaghetti and give a quick and good stir. Add the salmon steak, slightly fry for a while until cooked. Sprinkle with the spring onion, dish up and serve.

蒜蓉三文魚扒意粉

- dalam waktu 12 hari setelah melahirkan
- wanita setelah operasi caesar bisa makan
- apat makan setiap hari
- jika alergi terhadap bahan, tidak makan

Bahan:

1 organik ikan salmon (300 g)
1 sdk. bawang putih
(parut halus)
kering spageti
(jumlah tergantung pada selera individu)
1 bt. daun bawang
(potong kecil)
1000-2000 ml air
(tergantung pada petunjuk)
1/3 sdt. garam
(untuk memasak spageti)
1/2 sdt. minyak

Bumbu Pengasin:

2 sdt. air jeruk
1 sdt. minyak wijen
1/3 sdt. garam
sedikit lada hitam

Cara Membuat:

1. Ikan salmon cuci bersih. Pakai kertas dapur menyerap airnya.
2. Ikan salmon pakai bumbu pengasin, aduk rata, rendam 10 menit.
3. Pakai resep masakan atau selera individu spagheti memasaknya. Tiriskan.
4. Panaskan kewali dng. minyak, goreng ikan api besar masak 1 menit, goreng ikan 1 menit bolak-balik. Kasih bawang putih, api sedang, spaghetti aduk², goreng ikan sampai matang, kasih daun bawang. Jadilah.

Minced Beef Rice with Red Dates and Dried Tangerine Peel
Nasi Daging Sapi Cincang dan Tanggal, Kulit Jeruk Kering

- within 12 days after giving birth
- also suitable for women after caesarean
- take it once or twice a week
- women suffering from excessive internal heat should be cautious of taking it

Marinade:

1/3 tsp ground white pepper
1/3 tsp sugar
1/3 tsp salt
1/2 tsp sesame oil
1/2 tsp light soy sauce
1 tbsp rice wine
1/2 tbsp water

Ingredients:

1/2 tbsp shredded dried tangerine peel
200-300 g minced beef
10 red dates (pitted)
1/3 cup white rice
1-2 tsp chopped spring onion

Method:

1. Mix the marinade, add the minced beef and mix well. Leave for 30 minutes.
2. Finely dice the red dates. Soak the dried tangerine peel in water until soft, scrape off the pith and shred. Mix the red dates, tangerine peel and beef together and leave for 30 minutes.
3. Rinse the rice and put into a rice cooker to cook. When the water dries (rice about medium cooked), top the rice with the beef and steam until the rice is cooked through, referring to the signal of the rice cooker. Serve.

- -

- dalam waktu 12 hari setelah melahirkan
- wanita setelah operasi caesar bisa makan
- makan sekali atau dua kali seminggu
- jika penyakit panas internal yang berlebihan, berhati-hati

Bahan:

1/2 sdk. kulit jeruk kering
(potongan halus)
200-300 g daging sapi cincang
10 bj. tanggal (buang bijinya)
1/3 ukuran cangkir beras putih
1-2 sdt. daun bawang
(potong kecil)

Bumbu Pengasin:

1/3 sdt. lada putih
1/3 sdt. gula
1/3 sdt. garam
1/2 sdt. minyak wijen
1/2 sdt. kecap putih
1 sdk. arak beras
1/2 sdk. air

Cara Membuat:

1. Daging sapi cincang tambah bumbu pengasin, aduk rata, rendam 30 menit.
2. Tanggal potong kecil, kulit jeruk rendam pakai air dingin sampai empuk, potong potongan halus. Tanggal, daging sapi cincang, kulit jeruk aduk rata, rendam 30 menit.
3. Taruh beras ke dalam rice cooker, masak hingga airnya habis, masukkan daging sapi cincang, terus masak sampai matang nasinya. Jadilah.

陳皮免治牛肉紅棗飯

Swamp Eel Rice with Ginger Juice and Turmeric
Nasi Belut dan Sari Jahe, Bubuk Kunir

- to be taken within 12 days after giving natural childbirth
- to be taken after 12 days for women delivering by caesarean
- take it once or twice a week

- For those who are afraid of spicy food or easily suffer from excessive internal heat, the amount of ginger can be reduced to suit individual's physical condition, but it cannot be skipped.
- Those having an allergy to alcohol can reduce the quantity of glutinous rice wine to 1 tbsp.

Ingredients:

1/3 bowl ginger juice
300 g swamp eel
3 tbsp glutinous rice wine
1/3 cup white rice
1/2 tbsp shredded ginger
1/2 tsp oil

Marinade:

1/3 tsp sugar
1/3 tsp salt
1/2 tsp sesame oil
1/2 tsp light soy sauce
1/2 tsp turmeric powder

Method:

1. Cut open the swamp eel and rinse. Rub the inside and the outside with coarse salt to remove mucus. Rinse, remove the bone and then rinse again to remove the tiny bones. Cut into 2-inch long sections, combine with the mixed marinade and leave for 30 minutes.
2. Rinse the rice and cook in a rice cooker.
3. Heat oil in a wok. Sauté the ginger until aromatic. Put in the swamp eel and stir-fry until half cooked. Sprinkle with ginger juice. Turn to low heat, sprinkle with the wine and cook until the wine evaporates. Take out and set aside.
4. When the water in the rice cooker dries, top the rice with the swamp eel and continue to cook until the rice is done, referring to the signal of the rice cooker. Serve.

- untuk memberikan persalinan alami, makan dalam waktu 12 hari setelah melahirkan
- untuk wanita setelah operasi caesar, makan setelah 12 hari setelah melahirkan
- makan sekali atau dua kali seminggu
- jika penyakit panas internal yang berlebihan, pakai sedikit jahe
- jika alergi terhadap arak, gunakan 1 sdk. arak beras ketan

Bahan:

1/3 mangkuk sari jahe
300 g belut
3 sdk. arak beras ketan
1/3 ukuran cangkir beras putih
1/2 sdk. jahe (potongan halus)
1/2 sdt. minyak

Bumbu Pengasin:

1/3 sdt. gula
1/3 sdt. garam
1/2 sdt. minyak wijen
1/2 sdt. kecap putih
1/2 sdt. bubuk kunir

薑汁黃鱔薑黃焗飯

Cara Membuat:

1. Belut setelah dipotong baik, cuci bersih, siram garam, gosok² buang lender yg. dikulitnya. Belut cuci bersih, buang tulang, cuci bersih. Belut potong 2 inci, tambah bumbu pengasin, aduk rata, rendam 30 menit.
2. Beras cuci bersih, masukkan ke rice cooker.
3. Panaskan kewali dng. minyak, tumis jahe sampai wangi. Masukkan belut, aduk² sampai setengah matang. Masukkan sari jahe. Pakai api sedang masak, masukkan arak. Masak sampai menguap anggur.
4. Masak hingga airnya habis, masukkan belut. Terus masak sampai matang nasinya. Jadilah.

Pine Nut, Diced Vegetable and Minced Pork Fried Rice
Nasi Daging Babi Cincang dan Kacang Pinus, Choy Sum

- within 12 days after giving birth
- substitute the ginger for shallot for women after caesarean
- can be taken every day
- not suitable for those who are allergic to pine nuts, substitute pine nuts for other nuts

Ingredients:

1 tbsp pine nuts
10 Choy Sum stalks
200 g minced pork
1-1.5 bowls cooked white rice
2 tsp chopped spring onion
1 tsp grapeseed oil
3 ginger slices (grated)

Marinade:

1/3 tsp ground white pepper
1/3 tsp sugar
1/3 tsp salt
1/2 tsp sesame oil
1/2 tsp light soy sauce
1 tsp rice wine
2 tsp water

Method:

1. Stir-fry the pine nuts in a dry wok until fragrant.
2. Rinse and dice the Choy Sum stalks. Set aside.
3. Pinch the rice to loosen the grains.
4. Mix the marinade, add to the minced pork, mix well and leave for 30 minutes. Heat oil in a wok and stir-fry the pork until medium well. Remove and set aside.
5. Heat oil in the wok and sauté the ginger until fragrant. Put in the rice and stir-fry over high heat for 3 minutes. Reduce to medium heat. Add the minced pork and Choy Sum and stir-fry until cooked through. Sprinkle with the spring onion and give a good stir-fry. Dish up. Sprinkle the pine nuts on top before serving.

- dalam waktu 12 hari setelah melahirkan
- wanita setelah operasi caesar, pengganti jahe untuk bawang merah
- dapat makan setiap hari
- jika alergi terhadap kacang pinus, menggunakan kacang lainnya

Bahan:

1 sdk. kacang pinus
10 batang Choy Sum
200 g daging babi cincang
1-1.5 mangkuk masak beras putih
2 sdt. daun bawang (potong kecil)
1 sdt. minyak biji anggur
3 lb. jahe (parut halus)

Bumbu Pengasin:

1/3 sdt. lada putih
1/3 sdt. gula
1/3 sdt. garam
1/2 sdt. minyak wijen
1/2 sdt. kecap putih
1 sdt. arak beras
2 sdt. air

Cara Membuat:

1. Kacang pinus tumis tanpa minyak.
2. Batang Choy Sum cuci bersih, potong kecil.
3. Nasi melonggarkan sampai biji-bijian.
4. Daging babi cincang tambah bumbu pengasin, aduk rata, rendam 30 menit. Panaskan kewali dng. minyak, masukkan daging babi cincang, aduk^2.
5. Panaskan kewali dng. minyak, tumis jahe sampai wangi. Masukkan masak beras putih, aduk^2 3 menit. Masukkan daging babi cincang dan Choy Sum, aduk^2 sampai matang. Masukkan daun bawang, aduk^2. Masukkan kacang pinus. Jadilah.

松子菜粒肉碎炒飯

Red Rice with Chestnuts, Shiitake Mushrooms and Frog Leg
Beras Merah Kaki Katak dan Sarangan, Jamur Wangi

- within 12 days after giving birth
- can be taken every day
- do not take it if you are allergic to the ingredients
- substitute the ginger for dried tangerine peel for women after caesarean
- good to be treated as the meal taken between 3 am to 5 am before dawn

Ingredients:

10 shelled chestnuts
4 dried shiitake mushrooms
5 pairs Chinese edible frog legs
3-5 dried cloud ear fungus
1/3 cup red rice
1 tbsp shredded ginger

Marinade:

1/3 tsp ground white pepper
1/3 tsp sugar
1/3 tsp salt
1/2 tsp sesame oil
1/2 tsp light soy sauce
1 tsp rice wine

Method:

1. Rinse the frog legs, wipe dry, stir in the mixed marinade and leave for 30 minutes.
2. Soak the dried cloud ear fungus in water until soft, remove the stalks, rinse and tear to bite-sized pieces. Soak the dried shiitake mushrooms in water until soft, rinse, remove the stalks and shred. Skin the chestnuts, rinse and coarsely crush.
3. Rinse the rice and put into a rice cooker. Add the chestnuts, shiitake mushrooms, cloud ear fungus and ginger to cook together. When the water dries (rice about medium cooked), put the frog legs on top and steam until the rice is cooked. Serve.

- dalam waktu 12 hari setelah melahirkan
- wanita setelah operasi caesar, pengganti jahe untuk kulit jeruk kering
- dapat makan setiap hari
- Ini nasi sebaiknya di makan jam 3-5 pagi
- jika alergi terhadap bahan, tidak makan

Bahan:

10 bj. daging sarangan
4 bj. jamur wangi kering
5 pairs kaki katak
3-5 kering jamur kuping
1/3 ukuran cangkir beras merah
1 sdk. jahe (potong panjang)

Bumbu Pengasin:

1/3 sdt. lada putih
1/3 sdt. gula
1/3 sdt. garam
1/2 sdt. minyak wijen
1/2 sdt. kecap putih
1 sdt. arak beras

栗子冬菇田雞腿紅米飯

Cara Membuat:

1. Kaki katak cuci bersih, tambah bumbu pengasin, aduk rata, rendam 30 menit.
2. Kering jamur kuping pakai air sampai empuk, petik yg., cuci bersih, sobek menjadi potongan kecil. Jamur wangi kering pakai air sampai empuk, petik yg., cuci bersih, potong potongan halus. Daging sarangan buang kulitnya, cuci bersih, potong kecil.
3. Taruh beras ke dalam rice cooker, kasih daging sarangan, kering jamur kuping, jamur wangi kering, jahe, masak hingga airnya habis, masukkan kaki katak, terus masak sampai matang nasinya. Jadilah.

Garlic Pork Sparerib Steamed Brown Rice with Preserved Cabbage
Beras Kasar Tulang Iga Babi dan Bawang Putih, Sayuran Diawetkan

- within 12 days after giving birth
- also suitable for women after caesarean
- can be taken every day
- soak the brown rice longer in water to make it softer if you have a poor digestion

Ingredients:

2-3 cloves garlic
250 g pork spareribs
1 tsp preserved cabbage
1/3 cup brown rice

Marinade:

1/3 tsp ground white pepper
1/3 tsp sugar
1/4 tsp salt
1/2 tsp caltrop starch
1/2 tsp light soy sauce
1/2 tsp sesame oil
1 tsp rice wine

Method:

1. Bash the garlic, remove the skin and finely chop.
2. Rinse the pork spareribs and cut into bite-sized pieces. Stir in the mixed marinade, garlic and preserved cabbage. Leave for 60 minutes.
3. Rinse the brown rice and soak in water for 2 hours. Put into a rice cooker to cook. When the water nearly dries, put in the spareribs and continue to cook until the rice is done, referring to the signal of the rice cooker. Leave for 5 minutes. Press the button to cook again until the signal shows it is done. Serve.

- dalam waktu 12 hari setelah melahirkan
- wanita setelah operasi caesar bisa makan
- dapat makan setiap hari
- rendam lebih banyak waktu beras kasar dalam air untuk membuatnya lembut

Bahan:

2-3 bj. bawang putih
250 g tulang iga babi
1 sdt. sayuran diawetkan
1/3 ukuran cangkir beras kasar

Bumbu Pengasin:

1/3 sdt. lada putih
1/3 sdt. gula
1/4 sdt. garam
1/2 sdt. pati caltrop
1/2 sdt. kecap putih
1/2 sdt. minyak wijen
1 sdt. arak beras

Cara Membuat:

1. Bawang putih menampar, menghilangkan kulit, cincang halus.
2. Tulang iga babi cuci bersih, potong kotak², tambah bumbu pengasin, bawang putih dan sayuran diawetkan, aduk rata, rendam 60 menit.
3. Beras kasar cuci bersih, pakai air rendam 2 jam. Taruh beras kasar ke dalam rice cooker, masak hingga airnya habis kasih tulang iga babi, masak sampai matang, masak 5 menit, sampai matang. Jadilah.

香蒜排骨冬菜蒸糙米飯

Healthy Fried Glutinous Rice with Cubed Meat
Sehat Goreng Beras Ketan dan Daging Potong Dadu

- within 12 days after giving birth
- also suitable for women after caesarean
- take it once or twice a week
- if you have an allergy to peanuts, skip them and use fried olive nuts instead

Ingredients:

1 extra lean preserved sausage
100 g roast lean pork
1-2 dried shiitake mushrooms
2 cloves garlic
1 tsp canola oil
1/3 cup glutinous rice
1/2 cup water
2 sprigs spring onion (chopped)
10 fried peanuts

Seasoning:

1/4 tsp dark soy sauce
1/4 tsp sugar
1/3 tsp caltrop starch
1/3 tsp light soy sauce
1/3 tsp sesame oil
1 tsp rice wine

Method:

1. Soak the glutinous rice in water for 4 hours. Soak the dried shiitake mushrooms in water until soft, remove the stalks and cut into cubes. Soak the preserved sausage in hot water to wash slightly and remove grease and then dice. Dice the roast lean pork.
2. Bash the garlic, remove the skin and finely chop.
3. Put a little oil in a non-stick pan and sauté the garlic until fragrant. Put in the preserved sausage and stir-fry until hot. Add the shiitake mushrooms and stir-fry until cooked. Dish up and set aside.
4. Put the drained glutinous rice in the same pan. Turn to medium heat and stir-fry quickly. Lay the glutinous rice on the pan as even as possible to let the rice heat through. Add water several times, about 1-2 tbsp each time, and give a good stir-fry. Repeat adding water until the glutinous rice is cooked through (about 25-35 minutes). Add the seasoning and stir-fry until the sauce dries. Put in the fried ingredients and roast lean pork and give a good stir-fry. Sprinkle with the spring onion and peanuts and mix well. Dish up and serve.

- dalam waktu 12 hari setelah melahirkan
- wanita setelah operasi caesar bisa makan
- makan sekali atau dua kali seminggu
- jika alergi terhadap kacang tanah, menggunakan kacang zaitun goreng

健康臘味生炒糯米飯

Bahan:

1 bt. extra lean dachshund
100 g dibakar daging kurus
1-2 bj. jamur wangi kering
2 bj. bawang putih
1 sdt. minyak canola
1/3 ukuran cangkir beras ketan
1/2 ukuran cangkir air
2 bt. daun bawang
(potong kecil)
10 bj. kacang tanah goreng

Bumbu:

1/4 sdt. kecap hitam
1/4 sdt. gula
1/3 sdt. pati caltrop
1/3 sdt. kecap putih
1/3 sdt. minyak wijen
1 sdt. arak beras

Cara Membuat:

1. Beras ketan pakai air rendam 4 jam. Jamur wangi kering pakai air sampai empuk, petik yg., potong kecil. Dachshund pakai air panas rendam, potong kecil. Dibakar daging kurus potong kecil.
2. Bawang putih menampar, menghilangkan kulit, cincang halus.
3. Panaskan kewali sedikit minyak, tumis bawang putih, goreng dachshund, goreng jamur wangi kering. Angkat.
4. Pakai wajan goreng (3) pakai beras ketan kering masak pakai api sedang aduk^2 sampai rata atasnya. Kasih air beberapa kali, sekali gunakan 1-2 sdk. air, aduk^2. Tambah air hingga masak (25-35 menit). Kasih bumbu aduk^2 sampai tidak ber-air. Kasih bahan goreng dan dibakar daging kurus, aduk^2. Taruh daun bawang dan kacang tanah. Jadilah.

Stewed Peking Cabbage with Italian Ham
Peking Kubis dan Italian Ham

- within 12 days after giving birth
- can be taken every day
- for women after caesarean, sauté the ginger and remove it before putting in the vegetable

Ingredients:

50 g Italian ham
300 g Peking cabbage
1 tsp walnut oil
1/4 tsp salt
3 cloves garlic
3 ginger slices
2 tsp caltrop starch
(mixed with 1 tbsp cold boiling water)
1/3 cup water

Method:

1. Rinse the Peking cabbage, remove the old outer layer and cut into bite-sized pieces.
2. Bash the garlic, remove the skin and finely chop.
3. Heat oil in a wok. Sauté the ginger and garlic until fragrant. Add the Peking cabbage and quickly stir-fry over high heat for 2 minutes. Add water, turn to low-medium heat and stew with a lid on for about 15-20 minutes, or until it is tender and cooked. Season with salt, stir in the caltrop starch mixture and mix in the ham. Serve.

- dalam waktu 12 hari setelah melahirkan
- wanita setelah operasi caesar, tumis jahe dan menghapusnya sebelum menempatkan dalam sayuran
- dapat makan setiap hari

Bahan:

50 g Italian ham
300 g Peking kubis
1 sdt. minyak kenari
1/4 sdt. garam
3 bj. bawang putih
3 lb. jahe
2 sdt. pati caltrop
 (kasih 1 sdk. air mendidih dingin, aduk^2)
1/3 ukuran cangkir air

意
大
利
火
腿
燜
津
白

Cara Membuat:

1. Peking kubis cuci bersih, buang daun tua, potong kotak^2.
2. Bawang putih menampar, menghilangkan kulit, cincang halus.
3. Panaskan kewali dng. minyak, tumis jahe dan bawang putih sampai wangi. Masukkan masak peking kubis, aduk^2 2 menit. Kasih air, tutup dengan tutupnya, pakai api sedang masak 15-20 menit. Lalu kasih garam, kasih campuran pati caltrop dan ham, aduk rata. Jadilah.

Stir-fried Spinach with Enoki Mushrooms and Qi Zi
Tumis Bayam dan Jamur Enoki, Qi Zi

- within 12 days after giving birth
- a vegetarian diet
- also suitable for women after caesarean
- can be taken every day

Ingredients:

200 g enoki mushrooms
1 tbsp grated garlic
12 Qi Zi
500 g spinach
2 ginger slices
1/2 tsp oil
1/3 cup water

Seasoning:

1/2 tsp light soy sauce
1/2 tsp caltrop starch
1 tsp sesame oil
1/3 tsp salt

Method:

1. Rinse the spinach, remove the root and cut into 3-inch long sections.
2. Cut away the root of the enoki mushrooms and rinse.
3. Heat oil in a wok. Sauté the ginger and garlic until fragrant. Add the spinach and enoki mushrooms and quickly stir-fry over high heat for 2 minutes. Turn down the heat, add water, cover with a lid and simmer for about 15 minutes until the spinach and mushrooms are cooked. Put in the mixed seasoning and Qi Zi and stir-fry for 3 minutes. Dish up and serve.

金菇杞子炒菠菜

- dalam waktu 12 hari setelah melahirkan
- wanita setelah operasi caesar bisa makan
- vegetarian bisa makan
- dapat makan setiap hari

Bahan:

200 g jamur enoki
1 sdk. bawang putih
(parut halus)
12 bj. Qi Zi
500 g bayam
2 lb. jahe
1/2 sdt. minyak
1/3 ukuran cangkir air

Bumbu:

1/2 sdt. kecap putih
1/2 sdt. pati caltrop
1 sdt. minyak wijen
1/3 sdt. garam

Cara Membuat:

1. Bayam cuci bersih, tidak mau akar, potong 3 inci.
2. Enoki mushrooms cuci bersih, tidak mau akar.
3. Panaskan kewali dng. minyak, tumis jahe dan bawang putih sampai wangi. Masukkan bayam dan enoki mushrooms, masak 2 menit. Kasih air, tutup dengan tutupnya, pakai api sedang masak 15 menit sampai matang. Kasih aduk rata bumbu dan Qi Zi, tumis 3 menit. Jadilah.

Stewed Pork Spareribs with Pickled Plums

Tulang Iga Babi dan Acar Prem

- within 12 days after giving birth
- can be taken every day
- also suitable for women after caesarean

Ingredients:

50 g organic cane sugar
3 large pickled plums
300 g pork spareribs
3 ginger slices
2 cloves garlic

Marinade:

1 tsp light soy sauce
1.5 tsp dark soy sauce
1/3 tsp ground white pepper
1/3 tsp caltrop starch
2 tsp rice wine

Method:

1. Pit the pickled plums and press into paste.
2. Rinse the pork spareribs, cut into pieces and blanch in boiling water for 2 minutes. Remove, drain and let cool. Add the mixed marinade and pickled plum paste. Put into a refrigerator and leave for 30 minutes.
3. Put all the ingredients into a pot. Bring to the boil over medium-high heat, turn down the heat and stew for 25 minutes. Serve.

- dalam waktu 12 hari setelah melahirkan
- wanita setelah operasi caesar bisa makan
- dapat makan setiap hari

Bahan:

50 g organik sukrosa
3 bj. acar prem (besar)
300 g tulang iga babi
3 lb. jahe
2 bj. bawang putih

Bumbu Pengasin:

1 sdt. kecap putih
1.5 sdt. kecap hitam
1/3 sdt. lada putih
1/3 sdt. pati caltrop
2 sdt. arak bera

Cara Membuat:

1. Acar prem buang biji, tekan sampai menjadi pasta.
2. Tulang iga babi cuci bersih, potong kotak^2, rebus dalam air mendidih 2 menit, angkat, tiriskan, tunggu sampai dingin. Kasih bumbu pengasin dan acar prem pasta. Taruh 30 menit di dalam kulkas.
3. Panci besar kasih air semua bahan tutup panci. Pakai api sedang-tinggi masak sampai mendidih, pakai api sedang masak 25 menit. Jadilah.

梅子排骨

Stir-fried Chicken Fillet with Shallots
Tumis Daging Ayam dan Bawang Merah

- within 12 days after giving birth
- do not take it if you have a disease caused by external factors
- take it twice or three times a week
- For women after caesarean, sauté the ginger and remove it before putting in the chicken.

Ingredients:

4 shallots
300 g fresh chicken fillet
3 ginger slices
3 cloves garlic
1.5 tsp grapeseed oil
2 tsp Shaoxing wine

Marinade:

1/2 tsp dark soy sauce
1/2 tsp light soy sauce
1/2 tsp rice wine
1/2 tsp sugar
1/3 tsp salt
1/3 tsp ground white pepper
1/3 tsp sesame oil
1 tsp caltrop starch
(mixed with a little cold boiling water)

Method:

1. Rinse and skin the chicken fillet, remove the tendons, pick out the fascia (thin membrane) and cut into bite-sized pieces.
2. Mix the marinade, combine with the chicken and leave for half an hour.
3. Bash the shallots and garlic, remove the skin and chop finely. Heat oil in a wok, sauté the shallots and garlic for 2 minutes. Add the ginger and stir-fry for 1 minute. Sprinkle with the wine and cover tightly with a lid for 20 seconds. Turn to medium-high heat and quickly stir-fry the chicken for 3 minutes. Adjust to medium heat and stir-fry the chicken until done. Dish up and serve.

乾葱炒雞柳

- dalam waktu 12 hari setelah melahirkan
- wanita setelah operasi caesar, tumis jahe dan menghapusnya sebelum menempatkan dalam ayam
- makan dua kali atau tiga kali seminggu
- jika penyakit yang disebabkan oleh faktor eksternal, tidak makan

Bahan:

4 bj. bawang merah
300 g segar daging ayam
3 lb. jahe
3 bj. bawang putih
1.5 sdt. minyak biji anggur
2 sdt. arak masak

Bumbu Pengasin:

1/2 sdt. kecap hitam
1/2 sdt. kecap putih
1/2 sdt. arak beras
1/2 sdt. gula
1/3 sdt. garam
1/3 sdt. lada putih
1/3 sdt. minyak wijen
1 sdt. pati caltrop
(kasih sedikit air mendidih dingin, aduk²)

Cara Membuat:

1. Daging ayam cuci bersih, buang kulit, buang urat daging dan jalur, potong kotak².
2. Ayam tambah bumbu pengasin, aduk rata, rendam 30 menit.
3. Bawang merah dan bawang putih menampar, menghilangkan kulit, cincang halus. Panaskan kewali dng. minyak, masukkan bawang merah dan bawang putih, masak 2 menit. Masukkan jahe, masak 1 menit. Kasih arak beras, tutup dengan tutupnya, masak 20 detik. Pakai api sedang-tinggi, masukkan ayam, masak 3 menit. Pakai api sedang masak sampai matang. Jadilah.

Braised Abalone and Shiitake Mushrooms
Tumis Kental Pauhi dan Jamur Wangi

- within 12 days after giving birth
- take it once or twice a week
- also suitable for women after caesarean
- do not take it if you have a disease caused by external factors

Ingredients:

3 large Japanese shiitake mushrooms
1 canned large abalone
1 sprig spring onion
2 cloves garlic
2 tbsp bonito sauce
1/2 bowl abalone water
1/2 tsp canola oil

Seasoning:

1 tbsp rice wine
1.5 tsp dark soy sauce
1.5 tsp light soy sauce
20 g rock sugar
1 tsp caltrop starch (mixed with 1.5 tbsp cold boiling water)

Method:

1. Rinse all the ingredients. Soak the shiitake mushrooms in water until soft and then remove the impurities and stalks. Soak again in cold boiling water for 1 hour and cut into bite-sized pieces. Reserve the mushroom soaking water.
2. Cut off the root of the spring onion, chop the green parts and section the white parts. Bash the garlic and remove the skin.
3. Slice the abalone and set aside.
4. Add some oil in a pot, sauté the garlic until fragrant and then stir-fry the white spring onion. When it smells great, add the shiitake mushrooms and stir-fry for 5 minutes. Pour in the mushroom soaking water, bonito sauce and abalone water. Bring to the boil, turn down the heat and simmer until the shiitake mushrooms are cooked and tender. Mix in the abalone and seasoning. Adjust to medium heat and cook until the sauce dries. Add the chopped spring onion and mix well. Serve.

- -

- dalam waktu 12 hari setelah melahirkan
- wanita setelah operasi caesar bisa makan
- makan sekali atau dua kali seminggu
- jika penyakit yang disebabkan oleh faktor eksternal, tidak makan

Bahan:

3 ek. jamur jepang (besar)
1 ek. kalengan pauhi (besar)
1 bt. daun bawang
2 bj. bawang putih
2 sdk. saus bonito
1/2 mangkuk pauhi air
1/2 sdt. minyak canola

Bumbu:

1 sdk. arak beras
1.5 sdt. kecap hitam
1.5 sdt. kecap putih
20 g gula batu
1 sdt. pati caltrop
(kasih 1.5 sdk.air mendidih dingin, aduk[2])

Cara Membuat:

1. Semua bahan cuci bersih. Jamur jepang rendam sampai empuk. Lalu cuci bersih buang tulangnya. Lalu pakai air dingin rendam 1 jam. Potong menurut selera, jangan buang airnya.
2. Daun bawang buang akarnya. Batangnya dipotong panjang. Daun potong kecil. Bawang putih menampar, menghilangkan kulit.
3. Pauhi potong tipis.
4. Wajan taruh minyak. Bawang putih tumis sampai wangi. Taruh batang putih daun bawang tumis sampai wangi. Taruh jamur jepang, aduk[2], masak 5 menit. Ambil air yang tidak dibuang dalam (1), taruh bonito sauce dan pauhi air, masak hingga air mendidih. Masak hingga matang menggunakan api sedang. Taruh pauhi dan bumbu masak hingga kering. Kasih daun bawang. Kedalamnya. Jadilah.

鮑魚炆冬菇

Fried Sole Fillet with Broccoli and Cheese
Ikan Lidah dan Brokoli Keju

- within 12 days after giving birth
- can be taken every day

- substitute the ginger for garlic for women after caesarean
- those who are allergic to the ingredients should be cautious of taking it

Ingredients:

300 g broccoli
1.5 tbsp shredded cheese
2 tbsp milk
1 sole fillet (about 300 g)
3 cloves garlic
1 tsp oil
1.5 tsp lemon juice
1/4 tsp salt

Marinade:

1/3 tsp salt
1 tsp light soy sauce
1 tsp sesame oil
1/3 tsp sugar
1/3 tsp ground white pepper
2 tsp rice wine
1 tsp caltrop starch
1.5 tsp lemon juice

Method:

1. Defrost the sole fillet in the lower chamber of a refrigerator and then rinse. Spread lemon juice on the sole fillet entirely and leave for 5 minutes to remove the unpleasant smell. Add the mixed marinade and leave for 30 minutes.
2. Pour the milk into a small pot. Add the cheese and salt and simmer until the cheese completely melts, stirring constantly to avoid sticking to the pot.
3. Bash the garlic, remove the skin and chop finely.
4. Rinse the broccoli, cut into bite-sized pieces and blanch in boiling water until cooked. Drain and set aside.
5. Heat oil in a wok, sauté the garlic until fragrant and then fry the sole fillet until both sides are cooked. Add the broccoli and give a good stir-fry. Dish up. Pour the cheese sauce on top. Serve.

- dalam waktu 12 hari setelah melahirkan
- wanita setelah operasi caesar, pengganti jahe untuk bawang putih
- dapat makan setiap hari
- jika alergi terhadap bahan, berhati-hati

西蘭花芝士煎龍脷柳

Bahan:

300 g brokoli
1.5 sdk. keju parut
2 sdk. susu
1 pt. ikan lidah (300 g)
3 bj. bawang putih
1 sdt. minyak
1.5 sdt. air jeruk
1/4 sdt. garam

Bumbu Pengasin:

1/3 sdt. garam
1 sdt. kecap putih
1 sdt. minyak wijen
1/3 sdt. gula
1/3 sdt. lada putih
2 sdt. arak beras
1 sdt. pati caltrop
1.5 sdt. air jeruk

Cara Membuat:

1. Menempatkan ikan di lemari es, bukan freezer, untuk defrost. Cuci bersih. Air jeruk diperas taruh di atas ikan, diamkan 5 menit. Aduk rata bumbu pengasin, menambah ikan, diamkan 30 menit.
2. Menuangkan susu ke pot, kasih keju dan garam. Masak sampai keju meleleh, terus diaduk rata.
3. Bawang putih menampar, menghilangkan kulit, cincang halus.
4. Brokoli cuci bersih, potong kotak². Rebus dalam air mendidih, angkat, tiriskan.
5. Panaskan kewali dng. minyak, tumis bawang putih sampai wangi. Masukkan ikan sampai matang, masukkan brokoli aduk², lalu saus keju. Jadilah.

Salted Egg Yolk, Minced Pork and Pumpkin Stir-fry
Tumis Labu Kuning dan Kuning Telur Asin, Daging Kurus

- within 12 days after giving birth
- also suitable for women after caesarean
- take it once or twice a week
- substitute pork for chicken or beef as you wish

Ingredients:

1 salted egg yolk
200 g minced lean pork
300 g pumpkin
3 cloves garlic
1/2 tsp oil
1/3 cup water
2 tbsp chopped spring onion

Marinade:

1/2 tsp light soy sauce
1/2 tsp caltrop starch
1 tsp rice wine
1 tsp sesame oil
1/3 tsp ground white pepper
2 tsp water

Method:

1. Skin and seed the pumpkin and then cut into bite-sized pieces.
2. Bash the garlic and remove the skin.
3. Combine the minced pork with the mixed marinade and leave for 30 minutes.
4. Crush the salted egg yolk, add a little water, mix well and set aside.
5. Heat oil in a wok, sauté the garlic until fragrant, add the minced pork and stir-fry until medium well. Remove and set aside.
6. Quickly stir-fry the pumpkin in the same wok for 3 minutes, add water, cover with a lid and simmer for 15-20 minutes. Add the minced pork and stir-fry until done. Put in the salted egg yolk and stir-fry until cooked. Sprinkle with the spring onion and give a good stir-fry. Dish up and serve.

- dalam waktu 12 hari setelah melahirkan
- wanita setelah operasi caesar bisa makan
- makan sekali atau dua kali seminggu
- pengganti daging kurus untuk ayam atau daging sapi seperti yang anda inginkan

Bahan:

1 ek. kuning telur asin
200 g daging kurus cincang
300 g labu kuning
3 bj. bawang putih
1/2 sdt. minyak
1/3 ukuran cangkir air
2 sdk. daun bawang (potong kecil)

Bumbu Pengasin:

1/2 sdt. kecap putih
1/2 sdt. pati caltrop
1 sdt. arak beras
1 sdt. minyak wijen
1/3 sdt. lada putih
2 sdt. air

Cara Membuat:

1. Kupas labu kuning, buang biji, potong kotak^2.
2. Bawang putih menampar, menghilangkan kulit.
3. Daging kurus aduk rata dng. bumbu pengasin, diamkan 30 menit.
4. Kuning telur asin menghancurkan, kasih sedikit air, aduk rata.
5. Panaskan kewali dng. minyak, tumis bawang putih sampai wangi, masukkan daging kurus, masak sampai setengah matang.
6. Bekas kewali tumis labu kuning 3 menit. Kasih air, tutup dengan tutupnya, pakai api sedang masak 15-20 menit. Masukkan daging kurus, masak sampai matang, masukkan kuning telur asin, masak sampai matang. Jadilah.

鹹蛋黃肉碎炒南瓜

E Jiao and Egg Sweet Soup
Sup Manis E Jiao dan Telur Ayam

- 12 days after giving birth
- take it once or twice a week
- better consult a doctor of Chinese medicine beforehand
- substitute the ginger for dried tangerine peel for women after caesarean
- do not take it if you have a disease caused by external factors

Ingredients:

38 g high quality E Jiao
(donkey-hide gelatin)
1 egg
30 g brown sugar
450-500 ml water
3 slices old ginger

Method:

1. Ground the E Jiao.
2. Pour water into a pot, add the ginger and bring to the boil over high heat. Turn down the heat, add the ground E Jiao and brown sugar, keep stirring and cook until the E Jiao dissolves. Add the egg, mix well and simmer until the egg is cooked. Serve.

阿膠蛋蜜

- setelah 12 hari setelah melahirkan
- wanita setelah operasi caesar, pengganti jahe untuk kulit jeruk kering
- makan sekali atau dua kali seminggu
- jika penyakit yang disebabkan oleh faktor eksternal, tidak makan
- lebih baik berkonsultasi dengan dokter anda

Bahan:

38 g E Jiao
1 bj. telur ayam
30 g gula merah
450-500 ml air
3 lb. jahe tua

Cara Membuat:

1. E Jiao hancurkan halus.
2. Air masukkan kedalam panci, masukkan jahe, masak sampai mendidih. Pakai api sedang, masukkan hancurkan halus E Jiao dan gula merah, terus diaduk rata, sampai E Jiao lumer. Masukkan telur ayam, aduk rata, sampai matang. Jadilah.

Old Ginger Milk Pudding
Pudding Jahe Tua dan Susu

- 12 days after giving birth
- not suitable for women after caesarean
- take it once or twice a week
- women suffering from excessive internal heat should be cautious of taking it

Ingredients:

1/3 bowl old ginger juice
300-400 ml premium quality fresh milk
1/2 to 1 tbsp organic raw cane sugar

Method:

1. Pour the milk into a pot and bring to the boil over medium heat. Add the sugar and cook until the sugar dissolves.
2. Put the old ginger juice into a bowl. Pour the milk swiftly into the ginger juice right after the milk is removed from heat.
3. Leave to let the milk set. Serve.

- -

- setelah 12 hari setelah melahirkan
- wanita setelah operasi caesar, tidak makan
- makan sekali atau dua kali seminggu
- jika penyakit panas internal yang berlebihan, berhati-hati

Bahan:

1/3 mangkuk sari jahe tua
300-400 ml segar susu
1/2 to 1 sdk. organik sukrosa

Cara Membuat:

1. Susu masukkan kedalam panci, pakai api sedang, masak sampai mendidih. Masukkan sukrosa, masak sampai lumer.
2. Tuang sari jahe ke dalam mangkuk. Tuang susu panas dengan cepat ke dalam sari jahe.
3. Campuran (2) sampai bentuk berubah seperti jelli. Jadilah.

老薑汁撞奶

Vegetarian Ginger Vinegar Soup
Vegetarian Sup Jahe Cuka

- 12 days after giving birth
- make sure that the wounds are healed before taking it for women after caesarean
- do not take it if you have a disease caused by external factors
- take only a little if you suffer from excessive internal heat
- for women who feel cold in the body, good to serve the vinegar with rice

- The amount of this recipe is enough for a woman after giving birth to take regularly for about 15-20 times. It is most beneficial to the body by taking a little vinegar every day.
- Prepare the soup one month before delivery. Bring it to the boil once a week to prevent the growth of bacteria. By putting a lattice bamboo mat on the bottom of the ceramic pot, the ingredients will not stick to the pot. Wash the lattice bamboo mat and slightly blanch it in boiling water before cooking.
- For strict vegetarians who do not eat eggs, cook it with 1.2 kg of soy beans to increase the intake of protein.

Ingredients:

1.2 kg gluten puff
5000 ml sweetened black vinegar
3 kg old ginger
15-20 eggs
500 ml black glutinous rice vinegar
300 g slab sugar
1 tbsp salt

Method:

1. Rinse the gluten puff, blanch in boiling water, drain, cut into pieces and then squeeze water out. Fry in a dry wok for 10 minutes, dish up and set aside.
2. Boil the eggs until cooked and remove the shell.
3. Bash the skinned ginger and cut into pieces. Add salt and leave for 1 hour, rinse and drain. Line the bottom of a ceramic pot with a lattice bamboo mat designed for braising food, add all the ingredients and pour in the vinegar to cover the ingredients. Bring to the boil over high heat, turn down the heat and simmer until the gluten puff is tender. Turn off heat and soak in the eggs while hot (for at least 5 hours).

- -

- setelah 12 hari setelah melahirkan
- wanita setelah operasi caesar, memastikan bahwa luka disembuhkan
- jika penyakit yang disebabkan oleh faktor eksternal, tidak makan
- jika penyakit panas internal yang berlebihan, makan sedikit
- wanita yang merasa dingin, melayani sup jahe cuka dan nasi

- Resep Ini, sup jahe cuka dapat di makan 15-20 kali makan. Setiap hari minum sedikit sup jahe cuka agar sehat.
- Siapkan sup satu bulan sebelum pengiriman. Masak sampai mendidih seminggu sekali untuk mencegah pertumbuhan bakteri. Dengan meletakkan tikar bambu di bagian bawah panci, bahan-bahan tidak akan menempel pada panci. Sebelum masak, tikar bambu cuci bersih, rebus dalam air mendidih.
- vegetarian yang ketat, pengganti telur ayam untuk 1.2 kg kacang kedelai untuk meningkatkan asupan protein

素食養生薑醋

Bahan:

1.2 kg gluten engah
5000 ml cuka manis
3 kg jahe tua (buang kulit)
15-20 bj. telur ayam
500 ml cuka beras ketan hitam
300 g gula slab
1 sdk. garam

Cara Membuat:

1. Gluten engah cuci bersih, rebus dalam air mendidih, angkat, tiriskan, potong kotak^2, pemerasan untuk membuang air. Gluten engah tumis 10 menit tanpa minyak. Angkat.
2. Rebus telur ayam, buang kulitnya.
3. Jahe tua potong kotak^2, aduk rata dng. garam, diamkan 1 jam, cuci bersih, tiriskan. Menempatkan tikar bambu di bagian bawah panci. Kasih semua tarah dan cuka, pakai api besar masak sampai mendidih. Pasai api sedang, masak sampai gluten engah empuk. Matikan api, kasih telur ayam, rendam selama sebagai minimal 5 jam. Jadilah.

Nuo Dao Gen, Fang Feng and Jujube Tea
Teh Jujube dan Nuo Dao Gen, Fang Feng

- within 12 days after giving birth
- take it every day or once every two days until it produces the desired result
- also suitable for women after caesarean
- if in doubt of your physical condition, consult a doctor of Chinese medicine beforehand

Ingredients:

11 g roasted Bei Qi
8 g Nuo Dao Gen
8 g Fu Xiao Mai
8 g Fang Feng
4 jujubes
1 small piece dried tangerine peel
slab sugar to taste (optional)
1800 ml water

Method:

1. Put all the herbal ingredients into a mesh colander, rinse for a while, add water and soak for 15 minutes.
2. Bring all the ingredients (except sugar) to the boil over high heat, turn to low-medium heat and cook for 40 minutes. Filter the tea and season with sugar. Serve.

- dalam waktu 12 hari setelah melahirkan
- wanita setelah operasi caesar bisa makan
- dapat makan setiap hari atau makan sekali setiap dua hari sampai menghasilkan hasil yang diinginkan
- jika hasilnya tidak diinginkan, konsultasikan dengan dokter anda

Bahan:

11 g panggang Bei Qi
8 g Nuo Dao Gen
8 g Fu Xiao Mai
8 g Fang Feng
4 bj. jujubes
pojokan kulit jeruk kering
sedikit gula slab (fakultatif)
1800 ml air

Cara Membuat:

1. Bahan obat taruh kedalam jala saringan, cuci bersih, pakai air rendam 15 menit.
2. Bahan (kecuali gula) pakai api besar masak sampai mendidih, pakai api sedang masak 40 menit, saring ampasnya, lalu gula rasa. Jadilah.

糯稻根防風大棗茶

Dried Longan Tea with Tian Ma, Bai Zhi and Shi Chang Pu
Teh Lengkeng Kering dan Tian Ma, Bai Zhi, Shi Chang Pu

- within 12 days after giving birth
- take it every day or once every two days until it produces the desired result
- also suitable for women after caesarean
- if in doubt of your physical condition, consult a doctor of Chinese medicine beforehand

Ingredients:

8 g Tian Ma
 (processed with ginger juice)
8 g Bai Zhi
8 g Shi Chang Pu
11 g dried longan aril
6 g Chuan Xiong
6 g roasted liquorice
1 small piece dried tangerine peel
slab sugar to taste (optional)
1800 ml water

Method:

1. Put all the herbal ingredients into a mesh colander, rinse for a while, add water and soak for 15 minutes.
2. Bring all the ingredients (except sugar) to the boil over high heat, turn to low-medium heat and cook for 40 minutes. Filter the tea and season with sugar. Serve.

- dalam waktu 12 hari setelah melahirkan
- wanita setelah operasi caesar bisa makan
- dapat makan setiap hari atau makan sekali setiap dua hari sampai menghasilkan hasil yang diinginkan
- jika hasilnya tidak diinginkan, konsultasikan dengan dokter anda

Bahan:

8 g Tian Ma (diproses dalam sari jahe)
8 g Bai Zhi
8 g Shi Chang Pu
11 g daging lengkeng kering
6 g Chuan Xiong
6 g panggang liquorice
pojokan kulit jeruk kering
sedikit gula slab (fakultatif)
1800 ml air

Cara Membuat:

1. Bahan obat taruh kedalam jala saringan, cuci bersih, pakai air rendam 15 menit.
2. Bahan (kecuali gula) pakai api besar masak sampai mendidih, pakai api sedang masak 40 menit, saring ampasnya, lalu gula rasa. Jadilah.

天麻白芷石菖蒲桂圓茶

Slimming Tea
Teh Slimming

- within 12 days after giving birth
- take it every day or once every two days until it produces the desired result
- also suitable for women after caesarean
- if in doubt of your physical condition, consult a doctor of Chinese medicine beforehand

Ingredients:

11 g Bei Qi
11 g Ze Xie
11 g rice beans
8 g Yun Ling Pi
8 g Chong Wei Zi
8 g Tai Zi Shen
4 g roasted liquorice
4 g dried tangerine peel
4 g Zi Su Mu
1800 ml water

Method:

1. Put all the herbal ingredients into a mesh colander, rinse for a while, add water and soak for 15 minutes.
2. Bring all the ingredients to the boil over high heat, turn to low-medium heat and cook for 40 minutes. Filter the tea and serve.

- dalam waktu 12 hari setelah melahirkan
- wanita setelah operasi caesar bisa makan
- dapat makan setiap hari atau makan sekali setiap dua hari sampai menghasilkan hasil yang diinginkan
- jika hasilnya tidak diinginkan, konsultasikan dengan dokter anda

Bahan:

11 g Bei Qi
11 g Ze Xie
11 g kacang beras
8 g Yun Ling Pi
8 g Chong Wei Zi
8 g Tai Zi Shen
4 g panggang liquorice
4 g kulit jeruk kering
4 g Zi Su Mu
1800 ml air

Cara Membuat:

1. Bahan obat taruh kedalam jala saringan, cuci bersih, pakai air rendam 15 menit.
2. Bahan pakai api besar masak sampai mendidih, pakai api sedang masak 40 menit, saring ampasnya. Jadilah.

利水去腫修身茶

Light Fermented Black Bean, Chinese Chives and Malted Barley Tea
Teh Taosi Hambar dan Cina Kucai, Barley Malt

- can be taken whenever the reduction of breast milk supply is necessary
- substitute the ginger for dried tangerine peel for women after caesarean
- take it separately for many times until the result is satisfactory
- if in doubt of your physical condition, consult a doctor of Chinese medicine beforehand

Ingredients:

38 g light fermented black beans
300 g Chinese chives
38 g fried malted barley
2000 ml water
1 ginger slice

Method:

1. Rinse the Chinese chives and cut into 1-inch long sections.
2. Put all the herbal ingredients into a mesh colander, rinse for a while, add water and soak for 15 minutes.
3. Bring all the ingredients to the boil over high heat, turn to low-medium heat and cook for 40 minutes. Filter the tea and serve.

- jika wanita tidak mau ada air susu, minum Ini resep
- wanita setelah operasi caesar, pengganti jahe untuk kulit jeruk kering
- minum secara terpisah untuk berkali-kali sampai hasilnya memuaskan
- jika hasilnya tidak diinginkan, konsultasikan dengan dokter anda

Bahan:

38 g taosi hambar
300 g Cina kucai
38 g goreng barley malt
2000 ml air
1 lb. jahe

Cara Membuat:

1. Cina kucai cuci bersih, potong 1 inci.
2. Bahan obat taruh kedalam jala saringan, cuci bersih, pakai air rendam 15 menit.
3. Bahan pakai api besar masak sampai mendidih, pakai api sedang masak 40 menit, saring ampasnya. Jadilah.

淡豆豉韭菜麥芽茶

Blood Stasis Removing Tea
Teh untuk Menghapus Stasis Darah

- 12 days after giving birth
- take it separately for several times within 1 day
- take another dose one day after

- If you tend to have uncontrollable bleeding, consult a doctor of Chinese medicine beforehand.

Ingredients:

8 g Ai Ye
8 g liquorice
8 g Su Mu
11 g haw
1 small piece dried tangerine peel
1000 ml water
2 ginger slices

Method:

1. Put all the herbal ingredients into a mesh colander, rinse for a while, add water and soak for 15 minutes.
2. Bring all the ingredients to the boil over high heat, turn to low-medium heat and cook for 40 minutes. Filter the tea and serve.

- setelah 12 hari setelah melahirkan
- meminumnya beberapa kali dalam 1 hari
- minum dosis lain satu hari setelah

- jika cenderung memiliki perdarahan yang tidak terkendali, konsultasikan dengan dokter

Bahan:

8 g Ai Ye
8 g liquorice
8 g Su Mu
11 g pagar keliling
pojokan kulit jeruk kering
1000 ml air
2 lb. jahe

清宮祛瘀茶

Cara Membuat:

1. Bahan obat taruh kedalam jala saringan, cuci bersih, pakai air rendam 15 menit.
2. Bahan pakai api besar masak sampai mendidih, pakai api sedang masak 40 menit, saring ampasnya. Jadilah.

　　註冊中醫師許懿清是柏林醫務中心創辦人，兼任專上教育學院及多間專業培訓機構的講師。曾任職註冊護士及助產士近 10 年，其後在香港浸會大學修讀全日制中醫學位，畢業後一直行醫執業至今近 15 年。

　　近年許醫師更擔任國際著名護膚品牌的代言人及品牌中醫顧問，盡全力發揮其最擅長的中醫皮膚治療護理於各個領域之中，亦為 NOW TV 的健康節目擔任主持。

　　許醫師最愛鑽研婦科、皮膚科、兒科、針灸，以及產前、產後及初生嬰兒的治療與護理，對醫治各種與容顏、身體調理有關的疾病尤有興趣；經常自行烹調不同的食療藥膳，為家人的健康打好基礎。迄今已出版了 15 本食療著作，介紹如何運用醫食同源的概念維護健康，解決各種身體問題，寓養生保健於美食當中。

　　許醫師是一位極受傳媒及城中名人歡迎的中醫師，她經常在各大傳媒撰寫專欄、獲邀接受電視台及報章雜誌的訪問，希望藉着推廣中醫藥知識，提高中醫藥的認受性。

歡迎加入 Forms Kitchen「滋味會」

登記成為「滋味會」會員
- 可收到最新的飲食資訊 •
- 書展 "驚喜電郵" 優惠 *
- 可優先參與 Forms Kitchen 舉辦之烹飪分享會 •
- 每月均抽出十位幸運會員，可獲精選書籍或禮品 •

* 幸運會員將會收到驚喜電郵，於書展期間享有額外購書優惠

- 您喜歡哪類飲食叢書？(可選多於 1 項)

☐ 中菜　☐ 西菜　☐ 點心　☐ 烘焙　☐ 湯飲　☐ 甜品　☐ 其他＿＿＿＿＿＿

- 您對哪類飲食題材感興趣，而坊間未有出版品提供，請說明：

＿＿＿＿＿＿＿＿＿＿＿＿＿＿＿＿＿＿＿＿＿＿＿＿＿＿＿＿＿＿＿＿＿＿＿

- 此書吸引您的原因是：(可選多於 1 項)

☐ 興趣　　　☐ 內容豐富　　☐ 封面吸引　　☐ 工作或生活需要
☐ 作者　　　☐ 價錢相宜　　☐ 其他＿＿＿＿＿＿＿＿＿＿＿＿＿＿

- 您從何途徑擁有此書？

☐ 書展　　　☐ 報攤 / 便利店　☐ 書店 (請列明：＿＿＿＿＿＿＿)
☐ 朋友贈予　☐ 購物贈品　　☐ 其他

- 您覺得此書的價格：

☐ 偏高　　　☐ 適中　　　　☐ 因為喜歡，價錢不拘

- 除食譜外，您喜歡閱讀哪類書籍？(可選多於 1 項)

☐ 玄學　　☐ 旅遊　　☐ 心靈勵志　☐ 健康美容　☐ 語言學習　☐ 小説
☐ 兒童圖書　☐ 家庭教育　☐ 商業創富　☐ 文學　　☐ 宗教
☐ 其他＿＿＿＿＿＿＿＿＿＿＿＿＿＿＿＿＿＿＿＿＿＿＿＿＿＿＿

- 您是否有興趣參加作者的烹飪分享活動？

☐ 有興趣　　　☐ 沒有興趣

- 哪位作者的烹飪分享活動您會有興趣參加？

＿＿＿＿＿＿＿＿＿＿＿＿＿＿＿＿＿＿＿＿＿＿＿＿＿＿＿＿＿＿＿＿＿

姓名：＿＿＿＿＿＿＿＿＿＿＿＿☐ 男 / ☐ 女　　☐ 單身 / ☐ 已婚

聯絡電話：＿＿＿＿＿＿＿＿＿　電郵：＿＿＿＿＿＿＿＿＿＿＿＿＿＿

地址：＿＿＿＿＿＿＿＿＿＿＿＿＿＿＿＿＿＿＿＿＿＿＿＿＿＿＿＿＿

年齡：☐ 20 歲或以下　　☐ 21-30 歲　　☐ 31-45 歲　　☐ 46 歲或以上

職業：☐ 文職　　☐ 主婦　　☐ 退休　　☐ 學生　　☐ 其他＿＿＿＿＿

填妥資料後可：
寄回：香港鰂魚涌英皇道 1065 號東達中心 1305 室「Forms Kitchen」收
或傳真至：(852) 2565 5539
或電郵至：info@wanlibk.com

* 請 ✔ 選以下適用的項目
☐ 我已閱讀並同意萬里機構出版有限公司訂立的《私隱政策》聲明 #　☐ 我希望定期收到新書及活動資訊
有關使用個人資料安排
您好！為配合《2012 年個人資料（私隱）（修訂）條例》（《修訂條例》）的實施，包括《2012 年個人資料（私隱）（修訂）
條例》中的第 2(b) 項，萬里機構出版有限公司（下稱 "本社"）希望閣下能充分了解本社使用個人資料的安排。
為與曾跟萬里機構出版有限公司接觸的人士及已招收的會員保持聯繫，並讓閣下了解本社的最新消息，包括新書推介、會員
活動邀請、推廣及折扣優惠訊息、問卷調查、其他文化資訊及收集意見等，本社會不時向各位發放相關信息。本社會使用您的
個人資料（包括姓名、電話、傳真、電郵及郵寄地址），來與您繼續保持聯繫。
除作上述用途外，本社將不會將閣下的個人資料以任何形式出售、租借及轉讓予任何人士或組織。

坐好月子

A Complete Guide to Postpartum Recovery

作者	Author
許懿清	Hoi I Cheng, Veronica
策劃/編輯	Project Editor
	Karen Kan · Emily Luk
攝影	Photographer
	Imagine Union
美術統籌及設計	Art Direction & Design
	Amelia Loh
設計	Design
	Man Lo
出版者	Publisher
	Forms Kitchen
香港鰂魚涌英皇道1065號東達中心1305室	Room 1305, Eastern Centre, 1065 King's Road, Quarry Bay, Hong Kong
電話	Tel: 2564 7511
傳真	Fax: 2565 5539
電郵	Email: info@wanlibk.com
網址	Web Site: http://www.wanlibk.com
	http://www.facebook.com/wanlibk
發行者	Distributor
香港聯合書刊物流有限公司	SUP Publishing Logistics (HK) Ltd.
香港新界大埔汀麗路36號	3/F., C&C Building, 36 Ting Lai Road,
中華商務印刷大廈3字樓	Tai Po, N.T., Hong Kong
電話	Tel: 2150 2100
傳真	Fax: 2407 3062
電郵	Email: info@suplogistics.com.hk
承印者	Printer
中華商務彩色印刷有限公司	C&C Offset Printing Co., Ltd.
出版日期	Publishing Date
二零一四年六月第一次印刷	First print in June 2014
二零一八年五月第二次印刷	Second print in May 2018

免責聲明

本出版物提供的資料僅供參考之用,如有任何懷疑,服用前請諮詢中醫師。對於與本出版物有關連的任何因由所引致的任何損失或損害,本出版社與作者概不負責。